FARM TRACTOR
CLASSICS

Lee Klancher

Voyageur Press

First published in 2008 by MBI Publishing Company and Voyageur Press, an imprint of MBI Publishing Company, 400 First Avenue North, Suite 300, Minneapolis, MN 55401 USA

Copyright © 2008 by Lee Klancher

All rights reserved. With the exception of quoting brief passages for the purposes of review, no part of this publication may be reproduced without prior written permission from the Publisher.

The information in this book is true and complete to the best of our knowledge. All recommendations are made without any guarantee on the part of the author or Publisher, who also disclaim any liability incurred in connection with the use of this data or specific details.

We recognize, further, that some words, model names, and designations mentioned herein are the property of the trademark holder. We use them for identification purposes only. This is not an official publication.

Voyageur Press titles are also available at discounts in bulk quantity for industrial or sales-promotional use. For details write to Special Sales Manager at MBI Publishing Company, 400 First Avenue North, Suite 300, Minneapolis, MN 55401 USA.

To find out more about our books, join us online at www.voyageurpress.com.

Library of Congress Cataloging-in-Publication Data
Klancher, Lee, 1966–
 Farm tractor classics / by Lee Klancher.
   p. cm.
 ISBN 978-0-7603-3236-8 (hb, plc)
 1. Farm tractors. I. Title.
 TL233.K555 2008
 631.3'72--dc22

      2008003622
ISBN-13: 978-0-7603-3236-8

Editor: Leah Noel
Designer: Mandy Iverson

Printed in China

About the Author
Born and raised on the banks of the Brill River in northern Wisconsin, Lee Klancher has had a lifelong passion for books. Lee has authored seven books, including *Farmall: The Golden Age 1924–54*, *The Tractor in the Pasture*, and *Farmall Tractors*. His other work includes the 2005, 2008, and 2009 Farmall calendars and more than a hundred magazine feature articles covering everything from a tour of post-Katrina New Orleans to photographing the Bolivian national phenomena known as Caravana. Lee's work has appeared in more than 20 books as well as a myriad of motorcycle and travel publications, including *Motorcycle Cruiser* and *Robb Report Motorcycling*. He lives in St. Paul, Minnesota.

Additional Photo Credits
On the cover: A Ford 8N High-Crop. Ford built more than 500,000 of these machines. *Lee Klancher*

On the frontispiece: The radiator grille and lights of an International T-6. *Lee Klancher*

On the title pages: The New Hart-Parr, one of the first "small tractors" produced by the company. *Lee Klancher*

On the back cover: An International 1468, a model that debuted in 1971. *Lee Klancher*

On pages 6-7: Photos from the *Voyageur Press Archives*
On pages 24-25: *Lee Klancher* (main image); *Voyageur Press Archives* (inset)
On pages 84-85: Photos by *Lee Klancher*
On pages 158-159: Photos by *Chester Peterson Jr.*
On pages 206-207: *Lee Klancher* (main image); *Chester Peterson Jr.* (inset)
On pages 262-263: *Chester Peterson Jr.* (main image); *Hans Halberstadt* (inset)
On pages 306-307: Photos by *Lee Klancher*
On pages 350-351: *Hans Halberstadt* (main image); *Lee Klancher* (inset)
On pages 382-383: *Chester Peterson Jr.* (main image); *Hans Halberstadt* (inset)

Contents

Chapter 1 **THE FIRST FARM TRACTORS** 6

Chapter 2 **JOHN DEERE** 24

Chapter 3 **INTERNATIONAL HARVESTER** 84

Chapter 4 **FORD** 158

Chapter 5 **ALLIS-CHALMERS** 206

Chapter 6 **OLIVER** 262

Chapter 7 **MINNEAPOLIS-MOLINE** 306

Chapter 8 **CASE** 350

Chapter 9 **MASSEY** 382

Index 398

Chapter 1
THE FIRST FARM TRACTORS

*A*t the beginning of the twentieth century, the effects of the Industrial Revolution made their way onto the farm. The rural lifestyle that built America was replaced with a steam-powered planet that transformed subsistence farming into a mechanized business.

Steam-powered "tractors" appeared on the farm in the 1850s, providing portable units to power threshers. Later in the decade, early gas tractors came on to the market as lumbering beasts smaller than the giant steamers, but still impractical for most farm tasks. As technology progressed, farm equipment manufacturers developed smaller and lighter tractors. During World War I, the Fordson and a few other small tractors finally gave the farmer a light, affordable machine. The tractor played a key role in the transformation of the farm from an independent, self-sufficient lifestyle to a business capable of supporting an increasingly urban, mechanized world.

The First **CASE** Gas Tractor Built in 1892.

John Deere Plow

Many tractor manufacturers started out building plows, cultivators, and other pieces of agricultural equipment. Building farm equipment was big business in the 1800s, and the shrewd men of the time built large, powerful companies such as J. I. Case, McCormick, Deering, and Deere & Company. These formidable, profitable companies had the resources to spend the millions of dollars in research and development that was necessary to develop tractors in the early 1900s. *Voyageur Press Archives*

Minnesota Steam Tractor
Engine power came to the farm in the form of a steam engine on wheels. These portable engines ran threshing machines and the like. Self-propelled steam engines designed for farm use first appeared for sale in the 1850s and were widely used by the 1870s. These machines were large and expensive, and they were owned only by wealthy farmers or businessmen who rented the machines out for threshing and other operations. Farm equipment manufacture was big business by 1870, generating more than $48 million each year.
Voyageur Press Archives

Plowing with Steam
Steam tractors were used for breaking new ground, as evidenced by this photo from MacLeod, Alberta. Note the tanker behind the tractor used to keep a steady supply of water available while in the middle of large fields.
Voyageur Press Archives

Early Gas Tractors
Gas-engined farm tractors appeared at the end of the nineteenth century, and the first examples were slightly smaller and lighter replacements for the giant steam traction engines. This early International Harvester Company (IHC) gas tractor is a 15-horsepower Type A operating an asphalt mixer in Chicago's Grant Park in the early 1910s. Like early steam engines, these massive machines were expensive and used mainly on large farms, at corporations, or for industrial work. *Voyageur Press Archives*

Waterloo Boy Tractor

The first successful gas traction engine designed for agricultural use was built by Iowan John Froelich in 1892. He used a Van Duzen gas engine on a Robinson chassis and constructed his own transmission. A year later, John Froelich incorporated the Waterloo Gasoline Engine Company and, 20 years later, he started building the Waterloo Boy line of tractors. In 1918, John Deere purchased Waterloo in order to compete with the rest of the agricultural implement industry. One of the things that most impressed John Deere executives was the Waterloo Boy line's economical, simple two-cylinder engines. *Hans Halberstadt*

Waterloo Boy N

Early gas tractors shined at working large open plots of ground, as demonstrated by the 1923 photo of a Waterloo Boy discing in Russia. Foreign markets were something that the early farm equipment giants served well, and Russia was one of the world's largest exporters of grain in 1900. *Voyageur Press Archives*

Demonstrations
Early twentieth-century tractor makers marketed their machines with field demonstrations. Ads and flyers were posted around town, and then the machinery was hauled in and put to work in front of the local farmers. This technique was used by most of the manufacturers, and it was particularly popular with Ford and the International Harvester Company while the two giants were battling for market share in the early 1920s. *Voyageur Press Archives*

Rumely OilPull

One of the great names of early steam and gas tractors, Advance-Rumely was eventually folded into the Allis-Chalmers brand. Rumely's roots lie with two German immigrant brothers, Meinrad and Jacob Rumely, who established a blacksmith shop in LaPorte, Indiana, in the mid-1850s. Their M. & J. Rumely Company began building portable steam engines in 1872 and built kerosene and gas tractors later in its history. The company merged with Advance to form the Advance-Rumely Company, and it was purchased by Allis-Chalmers in the 1930s.
Chester Peterson Jr.

World War I Tractors
World War I created a demand for improved farm equipment, and manufacturers cashed in on this with ads like the one on the right. For the unfortunate farmers who stretched themselves too thin with debt in order to buy the new equipment, the ensuing decade proved tough. When prices for wheat and other crops dropped 60 percent in the early 1930s, farmers carrying significant debts found themselves bankrupt. *Voyageur Press Archives*

The Fordson
When Henry Ford introduced the Fordson in 1917, the agricultural industry was forever changed. The Fordson was small, light, and cheap—something that farmers could afford and use. When sales didn't meet Henry's expectations, he dropped the price to under $400, a move that gave the dominant tractor maker of the time, the International Harvester Company, fits and established the farm tractor as the up-and-coming implement on the farm. The true replacement of the horse wouldn't occur until after World War II, however, as only about 30 percent of American farms had tractors in 1945. *Voyageur Press Archives*

Chapter 2
JOHN DEERE

*T*he world's dominant maker of agricultural equipment has humble roots. Deere & Company was founded as a plow maker and now is an international enterprise with a 2007 annual revenue of $24.1 billion—more than the gross national product (GNP) of 150 countries.

John Deere started out building steel plows out of his blacksmith shop and built a giant agricultural equipment company in the 1800s. His company began building farm tractors in the 1910s, and soon Deere & Company made a name for itself by offering reliable twin-cylinder tractors and implements. John Deere was the second-largest tractor manufacturer throughout most of the first half of the twentieth century. In 1960, it took the lead as the premier American tractor manufacturer when the New Generation tractors were introduced. The rest of the industry has been playing catch-up ever since.

John Deere Plows

Deere & Company was founded by a young man, John Deere, who was working a small blacksmith shop in Grand Detour, Illinois, miles away from his wife and family back in Rutland, Vermont. Deere discovered that the western farmers needed a self-cleaning plow. He built and marketed the plow in the 1830s and used that success to create a powerful agricultural equipment manufacturing company. The company's annual revenue exceeded $2 million by 1900. *Hans Halberstadt*

John Deere in the Late 1800s

In 1848, the railroad bypassed Grand Detour, and, as a result, John Deere moved his young company to Moline, Illinois, where he could ship in steel and send plows around the country. By 1869, Deere & Company was making more than 40,000 plows each year. *Voyageur Press Archives*

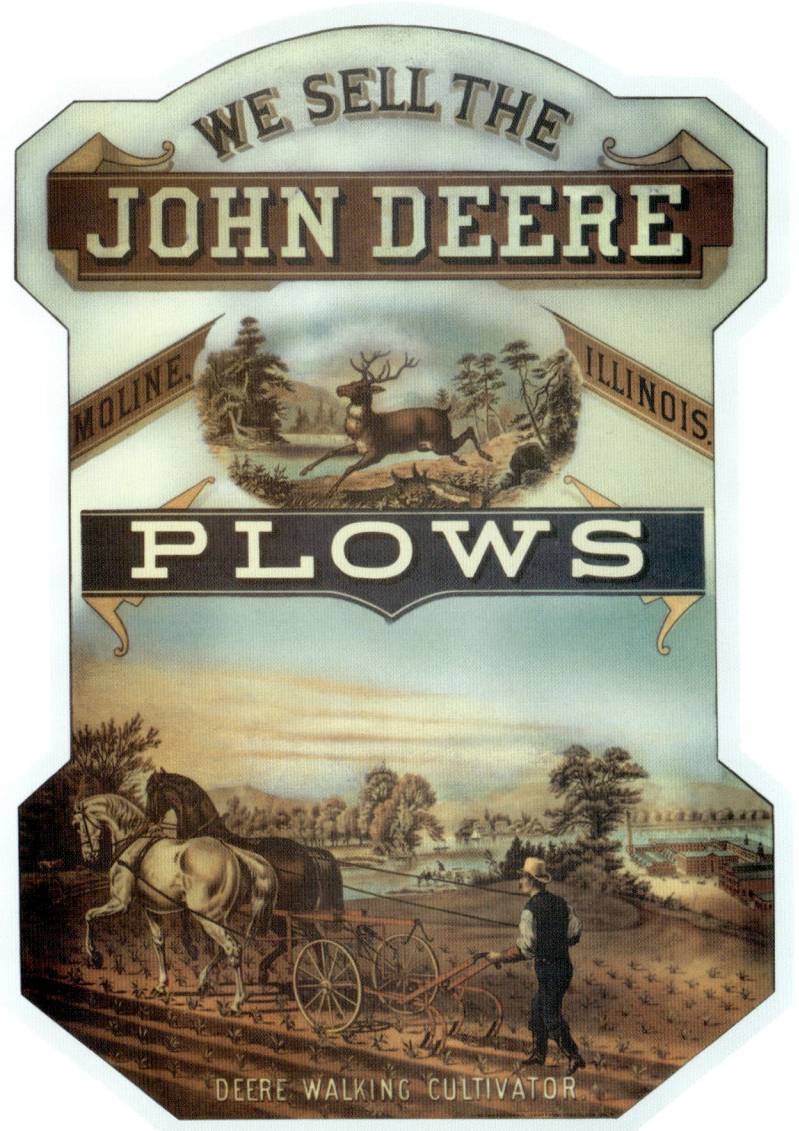

Growth by Acquisition
John Deere died in 1886, and his heirs grew the business by acquiring complementary agricultural manufacturers. By 1900, the company owned companies that built corn planters, wagons, hay tools, manure spreaders, grain drillers, and so on. The Deere & Mansur Company was founded by Charles Deere (John's second son) and Alvah Mansur to build corn planters. The company was rolled into Deere & Company in 1910, but the Deere & Mansur brand was used to badge a variety of John Deere implements.
Voyageur Press Archives

The Dain Tractor
In the early part of the twentieth century, Deere & Company needed to build a tractor to continue to be one of the dominant forces in the agricultural equipment business. It turned to board member Joe Dain, who developed this all-wheel-drive tractor for Deere. This machine was one of the first developed internally by the company, and it was a high-cost, high-technology machine. Dain died before the seventh example of his creation was built, and only a very few of them made it to market.
Voyageur Press Archives

Fordson Implements

The Dain Tractor program ended for a number of reasons. One was Dain's death in 1917. Deere & Company named another man to continue the work on the tractor, but the program was further derailed by wartime material shortages and Deere's efforts to build implements for the new-to-the-market Fordson. When Deere & Company bought Waterloo Boy, management chose to dedicate resources to market and develop Waterloo's two-cylinder line of tractors, and the Dain program officially came to an end. *Voyageur Press Archives*

Waterloo Boy Tractors

The most distinctive feature of early John Deere tractors is the rhythmic sound of the two-cylinder engines. The roots of this "pop-pop" rhythm came from the Waterloo Boy tractors, which Deere & Company purchased in 1918. The Waterloo Boy two-cylinder tractors were simpler and more economical than the more-sophisticated four-cylinder all-wheel-drive tractor developed for Deere by Joe Dain. *Voyageur Press Archives*

John Deere Model D
One of the gift horses handed to Deere & Company by the Waterloo Boy purchase was the Model D. The owners at Waterloo Boy didn't breathe a word about the machine until after the purchase was complete, and Deere & Company found it now had a modern, salable machine ready to go to market. The first widely sold tractor to bear the John Deere logo, the Model D was light, compact, and useful for the time, and it sold much better than the long-in-the-tooth Waterloo Boy Model N. *Hans Halberstadt (inset)/Voyageur Press Archives (opposite page)*

John Deere Model D

The Model D was a long-lived member of the Deere family. It was produced from 1923 to 1953. About 160,000 examples were built over that 30-year span. In 1924, a Model D cost $1,000. In 1953, the price was $2,124. *Hans Halberstadt*

John Deere Model D

The John Deere Model D was introduced late in 1923, a tough time in history to come out with a new model. The industry was recovering from the "tractor wars," initiated when Henry Ford was slashing prices on the Fordson. The companies with deep pockets matched Ford's low prices and lost money on their tractor lines in order to stay in business. The smaller companies went bankrupt. John Deere took a conservative route, and the tractor division barely survived. During 1921, John Deere sold only 79 of the 786 tractors the company manufactured. The John Deere Model D changed things for the company, and the tractor division became profitable by 1925. *Hans Halberstadt*

A Complete Line
The next model in John Deere's arsenal was the Model C (later known as the GP), a lighter, taller version of the Model D built with cultivation (and competing with IHC's popular Farmall) in mind. The model debuted in 1928, a time when Deere & Company was the second-largest agricultural equipment company in America. When tested at Nebraska, the early GP put out a rated 10 drawbar horsepower and 20 horsepower at the belt.
Voyageur Press Archives

Industrial Line
As Deere & Company grew the tractor line, the manufacturer produced a number of machines designed for purposes other than farming. The industrial line of tractors was designed for use in factories and highway departments, and it could even be ordered with these "turf" wheels and used to groom large expanses of grass and putting greens. The industrial models were painted orange, yellow, blue, and red.
Voyageur Press Archives

John Deere Model A

The Model A was introduced late in 1933 and was a more-developed version of the Model GP. The rear-tread width was adjustable, a helpful feature for farmers cultivating row crops of different widths. Note that the Model AO shown here is a wide-front tractor.
Hans Halberstadt

John Deere Model AO
Tractors built for special purposes, like this orchard model, are some of the most treasured by collectors. The orchard machines feature low-slung exhausts and cowlings over the controls and wheels to allow the tractor to slide through low-hanging orchard branches.
Hans Halberstadt

John Deere Model B
The John Deere Model B was introduced late in 1934 as a 1935 model, and it was a smaller version of the Model A. The Model B was produced until 1952 and was styled and redesigned over the years. This is an early example built in 1935. The Model B came equipped with a 4.25x5.25 bore-and-stoke engine that put out a rated 12 drawbar horsepower and 16 belt horsepower. The giveaway for very early examples is the fact that the front bolster is fastened to the frame with four bolts. Later unstyled Model Bs had bolsters fastened with eight bolts. *Lee Klancher*

John Deere Model B
The "General Purpose" logo was used on early John Deeres designed for cultivation. The experimental models used to develop the B were known as the Model HX and are extremely rare and valuable. The models built before 1939 are known as "unstyled" machines. The "styled" John Deeres have sleek sheet metal enclosing the front bolsters. *Lee Klancher*

John Deere Model B

In 1930, seven American companies offered complete lines of farm equipment: John Deere, the International Harvester Company, Allis-Chalmers, Minneapolis-Moline, Case, Oliver, and Massey-Harris. John Deere and International Harvester were the dominant makes then, and the two continue to vie for market share. Today, Deere is the dominant maker, while International has been absorbed into the giant farm equipment manufacturing corporation AGCO. *Lee Klancher*

John Deere Model B
Most early John Deere tractors were designed to run on kerosene, gasoline, or distillate. Hand clutches were used almost exclusively, and they were a bit awkward to operate. Engagement could be sudden, and the twin-cylinder engine's choppy power delivery didn't make taking off smoothly any easier. *Lee Klancher*

Lindeman Crawlers
When John Deere took an interest in crawlers, the tractor manufacturer turned to the Lindeman Company to mount tracks on its machines. The company owner, Jesse Lindeman of Yakima, Washington, converted John Deere tractors to be used by his customers in the area. The crawlers worked superbly in the sandy soil of the region.
Lee Klancher

John Deere Model BO Lindeman
Converting the early Model GPs to a crawler required extensive modifications. The Model B was much easier to convert, as the chassis design worked wonderfully with the Lindeman track system.
Lee Klancher

The Model L

The little John Deere L is one of the interesting departures in the John Deere line. The model was produced from 1937 to 1946, and the bodywork was smoothed and styled in 1939. The tractor uses a vertical twin-cylinder engine and a foot clutch, both atypical of the company's machinery at that time. The small tractor sold for about $500 in 1940 and cost nearly $600 in 1947. This L is working a No. 4 corn sheller. *Voyageur Press Archives*

A Stylish Legacy

In the 1930s, the industrial world discovered design, and the hot trend was sleekly styled machines ranging from refrigerators and living room furniture to automobiles and steam locomotives. Designers such as Russel Wright, Raymond Loewy, and Ferdinand Porsche were tapping into the growing desire for industrial products that integrated form and function. Tractors were no exception, and some of the best-known design firms in the world were called upon to bring style to the farm. John Deere turned to the design firm of Henry Dreyfuss for its styling exercise, and his emphasis on science and function as well as sleek lines created a line of timelessly attractive machines.
Randy Leffingwell

John Deere Model B
Henry Dreyfuss' most famous products include the *Twentieth Century Limited* locomotive, the Princess telephone, and the John Deere Model A and Model B. The Dreyfuss design firm applied its design touch to create the simply elegant Model B, which was the beginning of a long, cooperative relationship between the firm and Deere & Company. Dreyfuss design touches continued to grace John Deere tractors into the 1960s. *Voyageur Press Archives*

John Deere Model HWH
The Model G came out in 1938 as a larger general-purpose tractor with 20.70 rated drawbar horsepower and 31.44 drawbar horsepower. The Model H was introduced in 1939. The John Deere lineup up to that point was the Model A, Model B, and Model D. The Model H was introduced in 1939 to be a smaller general-purpose tractor. The HWH is a rare model that is very wide, allowing cultivation of crops that are set far apart.
Hans Halberstadt

Model M

Henry Ford was an innovator not only in the auto industry, but also in the world of tractors. First, his Fordson broke ground with its all-purpose design and rock-bottom price. Then Ford introduced his Model 9N in 1939, once again taking the agricultural world by storm. Ford's edge in 1939 was the Ferguson system, which allowed the little tractor to plow like a much larger machine. John Deere retaliated with the Model M, which was introduced in 1947 and produced until 1952. The M engine is an inline 100.5-cubic-inch twin, giving it a smoother exhaust note than that of most early John Deere twins.
Lee Klancher

Model R

Introduced early in 1949, the Model R was the first diesel tractor in the Deere lineup. The tractor also was the company's first to have an optional live power takeoff (PTO). The big two-cylinder 415.5-cubic-inch diesel was started with a small gas engine. Development of the model's highly regarded giant diesel engine took nearly a decade and demonstrated the company's dedication to producing high-quality machinery. *Voyageur Press Archives*

Model R

One of the biggest internal debates that went on during the development of the Model R was whether to make the tractor with a four-cylinder engine. John Deere's conservative, traditional management decided that the engine would be a two-cylinder, despite the considerable engineering challenges inherent in building a large-displacement diesel twin-cylinder engine. The Model R was built until 1954, and about 21,300 of them were created. This tractor was the end of the line for the John Deere letter series. The 40, 50, 60, 70, and 80 series of machines were launched in the mid-1950s. *Hans Halberstadt*

10 Series

Beginning in 1952, the letter series tractors were replaced by a new line of numbered models. Evolutionary changes to the model line added a live PTO, hydraulics, lights, and more power to the Deere line. The 40 replaced the Model M, the 50 replaced the Model B, the 60 replaced the A, and the 70 replaced the G. The new models were introduced over a three-year period. *Voyageur Press Archives*

Model 70

The Model 70 was the largest row-crop John Deere in the 10 Series. It came out in 1953 and was built until 1956. Four engines were available in the model—the gasoline, all-fuel, LP gas, and diesel versions. The 375.6-cubic-inch diesel-engine 70 set records for fuel economy when it was tested at the University of Nebraska. Note that tractors have been tested by the University of Nebraska since 1920, and those tests are now considered the most reliable source of information on horsepower, fuel economy, and other engine data for vintage farm tractors. *Voyageur Press Archives*

Model 320
The double-digit numbered series didn't last terribly long in the John Deere line, and the 20 Series was introduced in 1956. The smallest in the line was the 320, and its power was provided by a 100.5-cubic-inch twin-cylinder gasoline engine (a few were built with all-fuel engines, as well). In 1958, a Model 320 sold for $1,805. *Voyageur Press Archives*

Model 720 Diesel
The new color scheme on the 20 Series John Deere models came after consulting with the design firm of Henry Dreyfuss. The suggestion that came back was to make the side panels yellow to update the look, and that is precisely what John Deere did. *Lee Klancher*

Model 720 Diesel
The pony motor used to start the 720 diesel engine is a jewel of a motor. The little V-4 engine positively howls when it runs, and rebuilding the compact powerplant is an expensive proposition. Parts are hard to find and expensive—a distributor alone can set you back nearly $300. *Lee Klancher*

Model 830 Diesel

John Deere revised the 20 Series with updated sheet metal and new features for 1958 by creating the 30 Series. The big 830 was an evolution of the Model R diesel. The tractor featured hydraulic draft control, one of the most efficient agricultural diesel engines built, and an optional cab. By late in 1960, nearly 1.5 million John Deere two-cylinder tractors had been built and sold. *Chester Peterson Jr.*

The New Generation

On August 29, 1960, Deere & Company introduced a completely new line of farm tractors at a gala affair held in a stadium in Dallas, Texas. When the company finally decided to power its tractors with an engine possessing more than two cylinders, it also created an entirely new line of tractors. The engineers were allowed to start with a clean slate, and the results were revolutionary. The new line changed the industry and established John Deere as America's premier tractor manufacturer.
Voyageur Press Archives

Model 2010 High-Crop

The lines of the New Generation tractors were scrutinized carefully and designed to give the operator a clean view of the task at hand. The cleanly designed results speak to the John Deere objective—a design that is as functional as it is stylish—and to the scientific, cooperative approach taken by the Henry Dreyfuss design firm.
Voyageur Press Archives

Model 3010 and Model 4010

Inline four- and six-cylinder engines were used for the New Generation. The oil lines were cast into the block, making oil leaks less likely. The 3010 used a four-cylinder engine that was rated at 52.0 drawbar horsepower and 55.09 belt/PTO horsepower. The larger 4010 was powered by a six-cylinder 302-cubic-inch gas or 380-cubic-inch diesel engine. *Chester Peterson Jr.*

Model 1010 Industrial
John Deere offered industrial versions of the New Generation tractors, too. The Model 1010 was built from 1960 to 1965 and was available with a four-cylinder gas or diesel engine.
Chester Peterson Jr.

Model 8010/8020

The 8010/8020 was a John Deere–developed articulated four-wheel-drive tractor. The big machines put out more than 200 horsepower and had a retail price just over $30,000, making their appeal limited when they were introduced. A few of these machines were sold badged as 8010s and then recalled. The bulk of the 100 or so machines that were later sold were 8020s. *Chester Peterson Jr.*

Model 4020
The Model 4020 debuted in 1963, and production continued until 1972. The tractor produced nearly 100 horsepower at the PTO, and by 1972 the tractor's retail price exceeded $10,000. The 4020 was also available with front-wheel assist. *Chester Peterson Jr.*

Model 4955

The farm tractor became more sophisticated, expensive, and powerful over the years, and the John Deere 4955 built from 1988 to 1992 is a typical example of a modern row-crop machine. The inline turbocharged six-cylinder diesel engine provides 200 horsepower at the PTO. Speed is nearly infinitely adjustable due to a hydrostatic transmission, and the rear lift can raise more than five tons. *Voyageur Press Archives*

Model 9400

With 425 horsepower produced by a 765-cubic-inch six-cylinder turbodiesel, the 9400 has enough power to work big fields. As the ad touts, the big machine also has sophisticated lighting, so you can work late into the night. Produced from 1996 to 2002, the 9400 weighs more than 33,000 pounds, and its retail price was near that of an average home. *Voyageur Press Archives*

Chapter 3
INTERNATIONAL HARVESTER

When the companies of bitter rivals Cyrus McCormick and Charles Deering joined forces in 1902, the alliance of their powerful agricultural equipment dynasties helped create an agricultural equipment giant, the International Harvester Company. That company leveraged its massive market share advantage to become the world's largest farm equipment manufacturer for more than five decades.

The IHC claim to fame is the original Farmall tractor, now known as the Farmall Regular, a brilliantly engineered machine that is heralded as the first tractor to truly replace the horse. In the post–World War II boom, Farmall's H and M were the best-selling tractors of all time, with hundreds of thousands of the red machines selling each year. The company continued to make high-quality equipment, but struggles with labor and financing troubled the company in the 1960s and 1970s. In 1985, IHC's financial difficulties forced a sale to Tenneco (the owners of J. I. Case), and Case IH was formed. Today, Case IH lives on as part of Case New Holland (CNH).

The McCormick Dynasty

Cyrus McCormick is the best-known figure in International Harvester history, mainly because he is credited with inventing the reaper. The driven, fiery man created a dynasty, building his reaper company into one of the biggest players in American farm equipment manufacturing. McCormick died before the 1902 merger with Deering, and his widow, Nettie Fowler McCormick, remained a powerful figure in the company after he was gone. McCormick's descendants ran the company for a century. *Library of Congress*

Early IHC Tractors

The merger that created the International Harvester Company in 1902 gave the company a more than 80 percent market share. It owned all facets of agricultural equipment manufacturing, from twine production to companies that built binders, rakes, and tractors. Despite a massive lawsuit, and challenges from Henry Ford's products, the Chicago-based manufacturer was the dominant agricultural equipment company in America in the first half of the twentieth century. *Voyageur Press Archives*

McCormick-Deering 10-20

(Previous pages)
By the time the McCormick-Deering 10-20 was introduced in 1923, the IHC stranglehold on the tractor market was reduced to a death grip. That year, Henry Ford's factories cranked out more than 100,000 Fordsons, while International built only about 12,000 tractors. The McCormick-Deering 15-30 appeared in 1921, while the 10-20 rolled out of the factories in 1923. The machines were superbly engineered examples of an evolutionary design. The owner of this 1928 model recently rebuilt the engine and discovered that IHC offered a lifetime warranty on the engine bearings. The company honored that and gave him replacements for free. *Lee Klancher*

Farmall Regular

The IHC answer to the Ford challenge came not from the evolutionary McCormick-Deering line, but was the brainchild of brilliant engineer Bert R. Benjamin. Credited as the first successful all-purpose tractor, his revolutionary Farmall was the first to meld together the popular concepts of the day—a tricycle design, a four-cylinder engine, high clearance for cultivation, and multiple attachment points for a wide variety of implements. *Lee Klancher*

Farmall F-30

General-purpose tractors were the meat and potatoes of the tractor industry in the 1930s, but manufacturers continued developing larger machines capable of pulling a three- to five-bottom plow. Farmall's 15-30 was aging, and it was time to introduce a new big machine suited to breaking prairie and plowing the massive fields of the Great Plains. International's answer was the F-30, an enlarged Regular introduced in 1931 that was about 12 feet long (2 feet longer than the Regular) and weighed nearly two-and-a-half tons (nearly a ton heavier).
Lee Klancher

Farmall F-12

The F-12 was the fourth Farmall tractor. It was first produced in quantity in 1933, although experimentals and preproduction models were released early in 1932 and regular production models were released in October 1932. One of the F-12's most significant features was the elimination of the gear between the rear axle and rear wheel. By using larger rear wheels, the gear was no longer necessary.
Lee Klancher

Farmall F-12
The earliest F-12s used Waukesha engines, although Harvester quickly switched to an IHC unit. This 1933 F-12 is equipped with a 113-cubic-inch Waukesha engine. About 2,500 Waukesha-engine-equipped F-12s were built. *Lee Klancher*

Farmall W-12

The W-12 marked the beginning of a trend for International. While the W-30 bore little or no resemblance to the F-30, the W-12 and F-12 shared the same basic chassis. The result was reduced costs on everything from production to parts. The W-12 put out about 10 drawbar and 16 belt horsepower, similar numbers to the F-12. *Lee Klancher*

TracTracTor
The first International crawler was known simply as the TracTracTor, and it was built from 1928 to 1931. Note that the rear drive sprocket for the tracks is larger than the front. This was in part due to the fact that the steering clutches were housed in that rear sprocket, meaning service and wear were problems. *Voyageur Press Archives*

International I-12

The industrial version of the W-12, the I-12, rode on rubber tires and used a taller third gear than the W-12. The taller gearing and rubber tires made it easier to get around town or the company yard. Like the F-12 and W-12, the I-12 was upgraded to the I-14 in 1938, with a higher rpm rating that gave the tractor a bit more horsepower. *Lee Klancher*

Farmall M

The Letter Series is the most popular series of tractors ever built, combining cutting-edge technology with Raymond Loewy's classic design. The M was the fourth new Farmall introduced in 1939, and it was another incredibly popular Farmall. The largest of the line, the M produced a little more horsepower than that of International's discontinued big field workhorse, the McCormick-Deering 15-30. The M was rated to pull a three-bottom plow or turn the largest thresher or hammer mill.
Lee Klancher

Farmall H
More than 400,000 Farmall Hs were built from 1939 to 1954. This was the era in which the farm truly became mechanized, and you could rightly argue that the Farmall H brought power to the American farm. The Raymond Loewy–designed sheet metal is timelessly graceful and complements the lines on this HV, the high-crop version of the H. Along with Batman, penicillin, and Tupperware, the Farmall H is part of a select group of American icons from the 1940s that continues to thrive in the twenty-first century.
Lee Klancher

Farmall A

The Farmall A was the evolution of the original Farmall, as it was designed with cultivation in mind. The design gave the operator a clear view of the ground underneath the machine. In the end, the tractor's convenient size, economy, and ease of use that make it the popular machine it is even today. The Model B was similar to the A, with a different offset for the operator. The Model A was built from 1939 to 1954, while the B was made from 1940 to 1948.
Lee Klancher

International W-6

The Six Series tractors were lower-slung versions of the Farmall Model M, with the same engines as the M and sheet metal that was also designed by Raymond Loewy. The series was offered in orchard (O-6) and industrial (I-6) versions. The smaller Four Series (W-4, O-4, and I-4) used the Model H engine and chassis parts. The W-6 and its companion versions were built from 1940 to 1954.
Lee Klancher

International T-6

By the early 1940s, the International Harvester Company based whole lines of its machines on similar platforms, which cut production costs. Like the tractors, the crawlers wore sheet metal designed by Raymond Loewy. They also shared a number of parts with the Farmall M and the W-6 tractors. The T-6 crawler used the same engine as the tractors, but the crawler chassis was heavier, weighing about 7,000 pounds dry.
Lee Klancher

Farmall Cub

The Farmall Cub was introduced in the fall of 1945 and promoted heavily across the country. Despite the tractor's small size and mild 8-horsepower output, it had most of the features of its larger brethren—a four-cylinder engine, rubber tires, and a host of options that included power takeoff, a belt drive, a hydraulic lift, electric lights and starter, and a raft of International's trademark custom implements. The Cub was produced until 1964, making it one of IHC's longest-lived models.
Lee Klancher

Farmall C

The Farmall C was introduced in 1948 as a replacement for the Model B. The tractor was produced in respectable numbers, selling more than 20,000 a year in 1949 and 1950. If you include Super C production, more than 170,000 Model Cs were built from 1948 to 1954. The C was offered with Touch-Control, which was just another term for a hydraulic lift. In 1950, International celebrated the middle of the century with a limited run of models painted white and red.
Lee Klancher

International TD-14A

The original TD-14 was produced from 1939 to 1949 and was one of the larger crawlers in the lineup. This example is a TD-14A, which was built from 1949 to 1955. The machine is powered by a 461-cubic-inch four-cylinder engine and has six forward and two reverse speeds. *Lee Klancher*

Farmall Super C
In the early 1950s, IHC upgraded the Letter Series tractors to the "Supers," adding a mix of new transmissions, live PTO, Fast Hitches, and more horsepower to the lineup. The company built a Super A, Super C, Super H, and Super M. The Super C was built from 1951 to 1954, and it received new disc brakes and a larger cylinder bore for more horsepower.
Lee Klancher

Super M-TA
The last and best of the Letter Series line came in 1954, when IH rolled out the wonders of a fully independent PTO and torque amplification on the Super M-TA. A planetary gear set allowed the operator to switch speeds on the fly, a feature greatly appreciated when the load bogged the tractor down and a bit of extra power was needed. The tractor was produced at the height of IH's dominance of the tractor market, and nearly 30,000 Super M-TAs rolled out factory doors in 1954.
Lee Klancher

Super H-TA

If this sharply finished Farmall has you scratching your head, give yourself a couple of bonus points for historical knowledge. International did not ever build a Super H-TA, as the torque amplifier and live PTO of the "TA" variations were reserved for the Super M-TA and the W6-TA. This nicely done fake is comprised of parts from a number of different tractors, including a 300 and an H. Don't let the custom-made decals or the perfect restoration fool you—this model was never built by IHC. The superb finish on this tractor is pristine enough, however, to make you believe.
Lee Klancher

Farmall 400
The Hundred Series tractors reflected the first major revision of the International tractor line since 1939. The Hundred Series models were introduced late in 1954. The Farmall 400 was essentially a new skin and badge on the Super M-TA. Farmall 400s were produced from 1954 to 1956. The four-cylinder gas-engined versions could put out up to 48.1 drawbar horsepower and 52.4 PTO/belt horsepower.
Lee Klancher

Farmall 300 Super High-Crop

This Farmall 300 High-Crop has been fitted with an aftermarket kit that raises the tractor another 12 inches for use in sugar cane fields and with other crops that require extra clearance. Such machines are rare birds and are extremely valuable collector's items. The 300 was the evolution of the Model H, with an added independent PTO, Fast Hitch, and a Hydra-Touch hydraulic system. More than 29,000 300s were produced between 1954 and 1956.
Lee Klancher

International 600
The 600 was the improved version of the W-9, a big machine designed to break new ground. The design dated back to the 1940s, and the 600 was getting a little long in the tooth by the time of its production in 1956 and 1957. Production of the machine was fairly low, with only about 1,500 of them made.
Lee Klancher

International Wheatland 350
Only a few years after the Hundred Series was introduced, International came out with another "new" line. These tractors were not changed much outside of their new red-and-white paint scheme. The Wheatland 350s are a relatively rare version of the tractor produced for the Canadian market. They feature increased ground clearance, fixed-tread front axles, and optional independent PTOs.
Lee Klancher

International 350 High-Crop LP
The Holy Grail of tractor collecting is the high-crop LP model. This example is a 350, an improved version of the 300 that was built from 1956 to 1958. The 350 benefited from a little IHC tuning that squeezed a few more horsepower out of the engine. The 350 also had subtle improvements to the hydraulics and Fast Hitch. *Lee Klancher*

International 450 with Cummins Diesel

Like the 350, the 450 received subtle improvements over the 400. Traction control was added to the Fast Hitch system to provide steady draft control. The International 450 produced only 55 PTO horsepower, so aftermarket companies developed a number of kits that fitted more-powerful diesel engines. This rare IH is one of two development models used by Cummins. *Lee Klancher*

International 560 Diesel

The 560 is one of the most storied of the 1960 IHC line. Created to meet the growing demand for high-horsepower farm tractors, the 560 came out with a new inline six-cylinder that put out a stout-for-the-time 60 horsepower. The transmission and rear end weren't up to the power, and the tractor faced a recall of epic proportions. The company offered nearly unlimited overtime to employees in order to fix the problem, and stories persist to this day of people who bought homes using those overtime paychecks.
Lee Klancher

International 340

The fourth generation of the Hundred Series was the 40 Series (the third generation was the 30 Series, which was eventually followed by, and quite similar to, the 50 Series). To add to confusion, the 240 and 340 were introduced at about the same time as the new 460, 560, and 660. The 340 was built from 1958 to 1963. *Lee Klancher*

International Cub

The Cub was a mainstay in the IHC line, produced continuously (and changed very little) from 1947 to 1964. The little tractor was made for small-acreage farms, and more than 200,000 of the machines were built during its production run. The tractors are popular today due to their small size, their easily available parts, and the plethora of accessories that make them useful for cutting grass, plowing snow, and working small plots. This 1964 model sees use as a snowplow and is also equipped with a Cub 22 mower and a tandem disc.
Lee Klancher

International 2806

International brought out a number of significant tractors in the 1960s, but the 706 and 806 were the most successful. The high-horsepower tractors breathed new life into International and were billed as high-speed tractors. The transmission for the 706 and 806 was all-new. Considering the fiasco that occurred with the 560, International's last new high-horsepower model, the company took no chances with the 706 and 806.
Lee Klancher

International 2806

This rare bird is an industrial version of the venerable 806, which was the flagship of the IHC line when it was introduced in 1963. The high-horsepower tractors were available with front-wheel assist, and the front-wheel assist on this 2806 is a factory-installed option made by Coleman for IHC. The 2806 was available with both diesel and gasoline engines. This 1964 model came factory equipped with the 119-horsepower diesel engine, a Fast Hitch, two sets of hydraulic valves, and Torque Amplification (TA).
Lee Klancher

International 706

Front-wheel assist is one of the rarer features found on the International 706 line, and this 1965 706 was fitted with the front-wheel assist from an 806. In stock trim, the 232-cubic-inch six-cylinder diesel was good for 76 PTO horsepower. This 1965 Model 706 is fitted with an IH turbo that, along with the owner's tuning, has increased the PTO horsepower to an even 100. *Lee Klancher*

International 1256

This line began with the 1965 introduction of the 1206, the largest tractor in IHC's lineup. Horsepower in both diesel and gas models was just shy of 120, and the big tractors could be equipped with cabs outfitted with heat and air conditioning. In 1967, the 1206 received some relatively minor updates and became the 1256, which was produced until 1969. Note that the white vertical panel on the front of this 1969 model should have been painted red.
Lee Klancher

International 1468

The 1468 was an experiment in testosterone for IHC. The model came out in 1971 equipped with a DV-550 V-8 engine adapted from the truck line. The powerplant was added as much for marketing as it was for horsepower. The 549-cubic-inch V-8's 145-horsepower rating was no greater than the output of the 1466's six-cylinder engine, but the big chrome stacks and V-8 sound were designed to appeal to younger farmers. This example is a 1972 model. *Lee Klancher*

International 284

The 284 was a cutting-edge weapon in the war to sell tractors to small-acreage homeowners when it came out in 1976. It was a real tractor in miniature with an eight-speed transmission and either a Mazda gas or Nissan diesel engine. Power output was 26 PTO horsepower, and the slick tractor cost about $5,000 in its last year of production, 1984.
Lee Klancher

International 5288

Before International Harvester was purchased by Tenneco and became part of Case IH in 1984, the company produced a line of high-tech tractors known as the 30 Series and the 50 Series (along with a very few 70 Series articulated four-wheel-drives). The 5288, part of the 50 Series, is powered by a 466-cubic-inch engine rated for 186 PTO horsepower. The 5288 was introduced in 1981 and built until 1984. *Lee Klancher*

2007 Case IH (Steiger) STX 380
The relationship between Case IH and Steiger dates back to 1971, when International Harvester acquired a financial interest in the Fargo, North Dakota–based builder of high-horsepower four-wheel-drive tractors. The companies cooperated on the creation of the International 4366, and the relationship was made more formal when Case IH acquired Steiger in 1986. The first red Case IH Steiger models, the 9100 Series, appeared in 1988. The STX380 is an evolution of that partnership and features a 418-horsepower biodiesel-capable engine, 16-speed transmission, and 300 gallons of fuel capacity. Base price is about $200,000, with a nicely equipped model going out the door for more than $300,000.
Lee Klancher

Chapter 4
FORD

*I*n the first half of the twentieth century, Henry Ford was the agricultural equipment industry's corporate Godzilla, rising occasionally out of the depths to shake things up. His first strike was the Fordson, a cheap, useful, machine that bankrupted lesser tractor companies and forced the industry to make tractors smaller and more affordable. His next attack was the Ford-Ferguson 9N, which combined Harry Ferguson's revolutionary draft control system with a compact, modern tractor chassis. Tractor makers spent years struggling to find an alternative to the Ferguson three-point hitch, and eventually it became industry standard.

After an acrimonious split with Ferguson, the Ford Motor Company decided to become a full-fledged agricultural equipment manufacturer and, after a flurry of mergers and acquisitions, the brand lives on today.

Mak-a-Tractor

The earliest Ford "tractors" were Model Ts equipped with kits that lowered the overall gear ratio of the vehicle and added rear wheels with better traction for the fields. The conversions didn't make great farm tractors, but they were an economical way to put power on the farm. *Voyageur Press Archives*

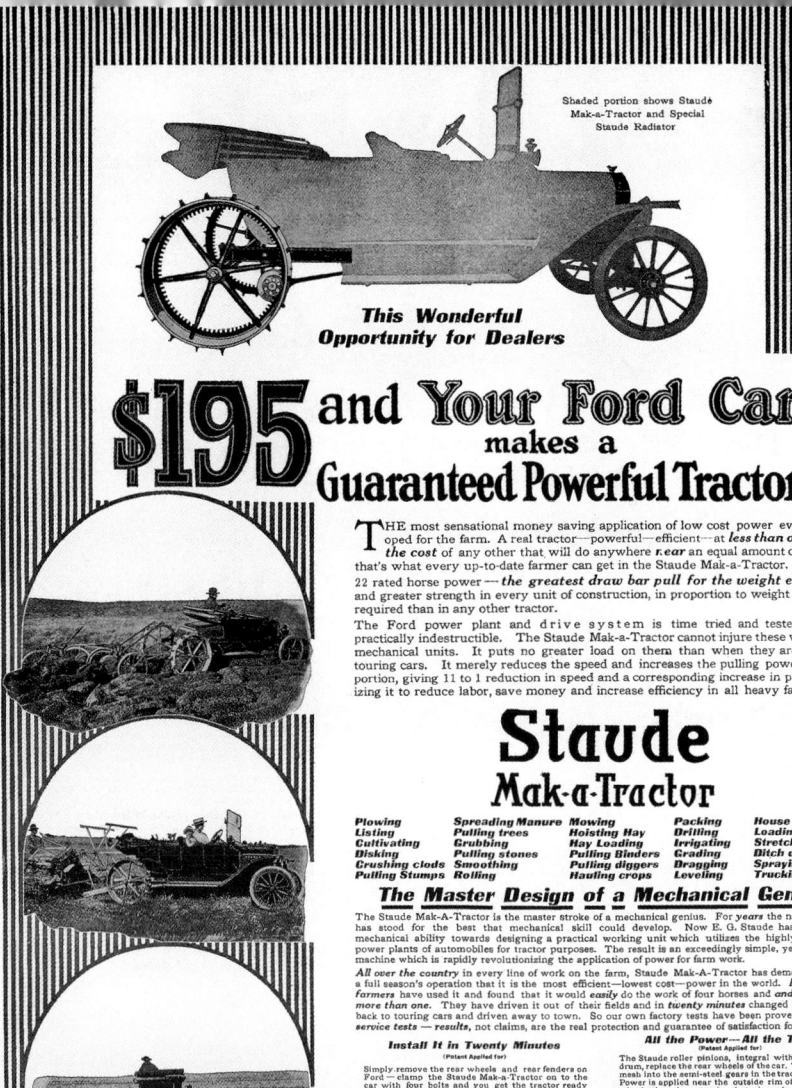

Shaded portion shows Staude Mak-a-Tractor and Special Staude Radiator

This Wonderful Opportunity for Dealers

$195 and Your Ford Car makes a Guaranteed Powerful Tractor

THE most sensational money saving application of low cost power ever developed for the farm. A real tractor—powerful—efficient—at *less than one-third the cost* of any other that will do anywhere near an equal amount of work—that's what every up-to-date farmer can get in the Staude Mak-a-Tractor.

22 rated horse power —*the greatest draw bar pull for the weight ever built* and greater strength in every unit of construction, in proportion to weight and work required than in any other tractor.

The Ford power plant and drive system is time tried and tested. It is practically indestructible. The Staude Mak-a-Tractor cannot injure these wonderful mechanical units. It puts no greater load on them than when they are used as touring cars. It merely reduces the speed and increases the pulling power in proportion, giving 11 to 1 reduction in speed and a corresponding increase in power utilizing it to reduce labor, save money and increase efficiency in all heavy farm work.

Staude Mak-a-Tractor

Plowing	Spreading Manure	Mowing	Packing	House moving
Listing	Pulling trees	Hoisting Hay	Drilling	Loading logs
Cultivating	Grubbing	Hay Loading	Irrigating	Stretching wire
Disking	Pulling stones	Pulling Binders	Grading	Ditch digging
Crushing clods	Smoothing	Pulling diggers	Dragging	Spraying
Pulling Stumps	Rolling	Hauling crops	Leveling	Trucking

The Master Design of a Mechanical Genius

The Staude Mak-A-Tractor is the master stroke of a mechanical genius. For *years* the name Staude has stood for the best that mechanical skill could develop. Now E. G. Staude has turned his mechanical ability towards designing a practical working unit which utilizes the highly developed power plants of automobiles for tractor purposes. The result is an exceedingly simple, yet wonderful machine which is rapidly revolutionizing the application of power for farm work.

All over the country in every line of work on the farm, Staude Mak-A-Tractor has demonstrated in a full season's operation that it is the most efficient—lowest cost—power in the world. *Everywhere farmers* have used it and found that it would *easily* do the work of four horses and *and it costs no more than one.* They have driven it out of their fields and in *twenty minutes* changed their Fords back to touring cars and driven away to town. So our own factory tests have been proven by *actual service tests — results,* not claims, are the real protection and guarantee of satisfaction for owners.

Install It in Twenty Minutes
(Patent Applied for)

Simply remove the rear wheels and rear fenders on Ford—clamp the Staude Mak-a-Tractor on to the car with four bolts and you get the tractor ready for use.

Leave the Mak-a-Tractor channel irons on if you want to when you change back to the touring car—they will not interfere with riding or appearance—or you can remove them in five minutes.

No holes to bore, nothing to change. All you need for tools is a wrench, a jack and a wheel puller. We furnish the puller—the work takes 20 minutes.

All the Power—All the Time
(Patent Applied for)

The Staude roller pinions, integral with the brake drum, replace the rear wheels of the car. The pinions mesh into the semi-steel gears in the tractor wheels. Power is applied near the outside rim of the tractor wheel—where is no torsional strain on hub or spokes.

And the Staude Mak-a-Tractor Axle is *back of the* car axle—an exclusive construction. The driving pinions *push* the tractor wheels down—no power is wasted—you get all the power all the time.

So wonderfully is the power applied that the Ford used with the Staude Mak-a-Tractor not only runs in high gear in the hardest work, but it actually starts on high.

The Automobile Plow

The Ford Motor Company started experimenting with agricultural vehicles in the early part of the century. The first of Henry Ford's significant experiments was the Automobile Plow, which was completed in 1905 and refined over the next few years. Ford also experimented with larger tractors before switching to a smaller design (which eventually became the Fordson). *Voyageur Press Archives*

The Other Ford Tractor

In the early 1910s, San Francisco promoter and all-around shyster William Ewing smelled money when he hired a shop clerk named Paul B. Ford. Ewing appointed Ford the director of a new company, the Ford Tractor Company, and quickly commissioned an independent firm to draw up (or steal—the records are not entirely clear) a tractor design. Ewing sold a few of these poorly built "Ford" tractors in the middle of the decade before the company went bankrupt. Several lawsuits were filed against Ewing and his cronies, none of which were found to contain sufficient evidence to prosecute. *Voyageur Press Archives*

THE FORD TRACTOR

See it on the Main Floor (lobby) West Hotel During Implement Dealers' Convention, Jan. 11-13

During the last half of the year 1915, Ford Tractors were shipped to purchasers in various parts of the following states: **Minnesota, Wisconsin, Iowa, North Dakota, South Dakota, Montana, Indiana, Illinois, Kansas, Oklahoma, Missouri, New Mexico, Texas and New York.** Our many Dealers in these States are "lining up" for the big 1916 business, which is believed will be the biggest tractor year in history. **Make your arrangements now for the Ford Tractor agency in your locality.**

THE FORD TRACTOR COMPANY

General Office and Sales Rooms: Ford Tractor Building, 2619 to 2627 University Avenue S. E.,
MINNEAPOLIS, MINN.

The Overseas Fords

The first customer for the Fordson was the British Ministry of Munitions, who needed the tractors during World War I. Henry chose to avoid the controversy created with the Ford tractor name by William Ewing, and his first farm tractors shipped over to Europe were discreetly badged "Henry Ford & Son." *Voyageur Press Archives*

The Fordson

In 1918, the Fordson made its American market debut. Henry's tractor was loud, it could easily flip over backwards, and the transmission heat blasted a farmer's backside under heavy loads. Even so, the machine was cheap, effective, and appeared on the market just as World War I created a demand for increased productivity on the farm. In the eight years after it was introduced, the Fordson literally dominated the tractor market, outselling the more-established line sold by International Harvester by a two-to-one margin in the early years. From 1922 to 1926, Ford tractors comprised more than 60 percent of all tractor sales in America. In 1923, Ford sold about 100,000 Fordsons, and International sold just over 12,000. Total tractor sales that year were only about 132,000, meaning that Ford owned 76 percent of the market that year. *Hans Halberstadt*

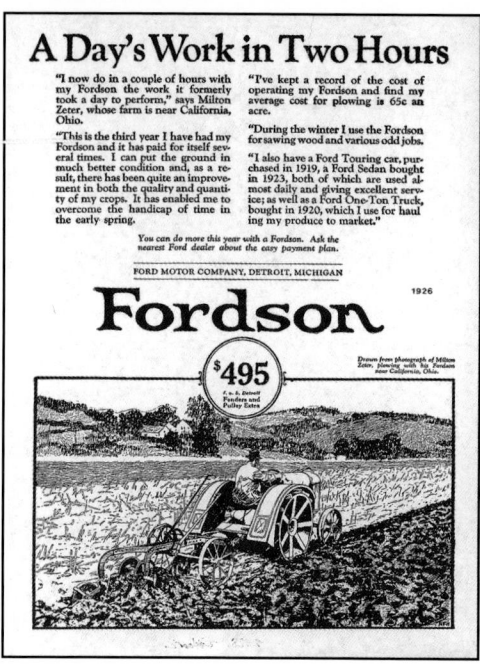

Tractor Wars

When Henry Ford slashed the $700 price of the Fordson to $395 in 1922, he was selling machinery below cost. International Harvester matched Ford's cuts, and the two manufacturers' money-losing battle for market dominance in the mid-1920s is known as the tractor wars. In 1928, Ford stopped producing Fordsons in the United States and focused his efforts on selling tractors overseas. *Voyageur Press Archives*

The European Fordson

When production of the Fordson stopped in the United States, the focus for Ford shifted to building and selling the tractors in the plants in the United Kingdom and Ireland. Several models were made over the years, including the Fordson N, the Fordson F, and the All-Around. *Hans Halberstadt*

The European Fordson

The European Fordsons were initially built at a plant in Cork, Ireland. Production moved to the Ford plant in Dagenham, England, in the early 1930s. Early models sported blue paint with orange trim. After November 1937, the Fordsons were painted orange with black trim. *Hans Halberstadt*

Fordsons in Canada

The European-built Fordsons were exported to North America in limited quantities, with less than 5 percent of the production run coming to the United States. This Canadian Fordson brochure is most likely from the late 1930s, since it contains options for pneumatic tires, which became available in 1935. *Voyageur Press Archives*

Early Prototype

In 1931, Henry Ford had his engineers experimenting with tractor development. One of the prototypes was this machine, which was created by melding a flathead V-8 with a number of Ford truck parts, including the steel frame rails and grille. Late in the 1930s, Ford also paraded an awkward little three-wheel prototype in front of the best and brightest in the industry. Whether that was a clever ruse or evidence that Ford didn't yet have his next big thing in agriculture, it meant he was more than interested in the opportunity that Harry Ferguson presented him with a few years later. *Chester Peterson Jr.*

Flathead Power

The engine for this prototype was a Ford flathead V-8, and part of Henry's mandate was that initial prototypes be built around this engine. This prototype is an anomaly, and factory records are scarce. The tractor survived because it was used at Ford's Fair Lane Estate. *Chester Peterson Jr.*

The Handshake
The Ford 9N was created when Harry Ferguson (seated) demonstrated his revolutionary draft system to Henry Ford I (in glasses). Ferguson scraped up every last dime to fly a demonstrator tractor equipped with his system from England to Henry's estate, Fair Lane, near Dearborn, Michigan. Henry spent five days observing, prodding, and parading the tractor and its system in front of company executives. At the end of the fifth day, Ford asked for a table and chairs to be set out, and the two sat down to make a deal. On November 8, 1938, the men shook hands and agreed to go in business together and build a farm tractor.
Voyageur Press Archives

1939 Ford-Ferguson 9N

Ferguson and Ford's handshake resulted in a flurry of design activity at Ford's River Rouge plant. Most of the design work centered around transforming the Ferguson design to work for American farmers and to make it look like a Ford. By June 29, 1939, a running prototype was demonstrated to a group of 460 select Ford guests at the Dearborn Inn. Early 9Ns were built with polished cast-aluminum hoods. *Chester Peterson Jr.*

The Ford-Ferguson 9N

In 1939, Ford got back into the tractor business and threw another monkey into the industry's plans. The Ford-Ferguson 9N was introduced, and, with the help of the Ferguson system (better known as the three-point hitch), the diminutive Ford could do twice the work of a Farmall F-30 at half the price. This tune was familiar to Ford fans, and IHC's profit share was cut into once again. *Chester Peterson Jr.*

Ford-Ferguson Brochure
This brochure touts the advantages of the new Ford-Ferguson 9N, particularly the fact that the Ferguson three-point hitch allowed the small machine to plow like a much larger tractor. Ferguson gambled everything he had on the partnership, and it paid off for him. *Voyageur Press Archives*

The Ferguson System
The Ferguson system, now known as a three-point hitch, provided better draft control than previous plow systems and eliminated the problem of a tractor rearing up and flipping when a plow hit a rock or other solid object. The three-point hitch was used exclusively by Ford and Ferguson until the patent expired, at which time the three-point hitch quickly became the industry standard. It is still used on modern farm tractors. *Chester Peterson Jr.*

The Ford 9N

The Ford N Series consists of the Ford 9N, 2N, and 8N. The three models are very similar, with minor upgrades to each line. The 9N was produced from 1939 to 1941, the 2N from 1942 to 1947, and the 8N from 1948 to 1952. Note that the tractors' names were derived from the last digit of the year the machines were introduced. Henry Ford and Harry Ferguson's handshake agreement led to an acrimonious relationship in which neither partner was entirely satisfied. In 1947, Ford and Ferguson bitterly parted ways. Ford revised its hitch, and the 2N no longer bore the Ferguson name. Ferguson built his own tractors and hired lawyers to go after Ford. *Chester Peterson Jr.*

Ford 8N High-Crop
The 8N no longer bore any tip of the hat to Ferguson, which is not to say that Ford was free from its obligation to the company. Ferguson filed a lawsuit against Ford, and the two settled out of court, with Ford paying just over $9 million to Ferguson. The 8N is one of the most common models of tractors. Ford built more than 500,000 of the machines. An aftermarket kit raised the height of this tractor with lengthened front spindles and larger rear wheels.
Lee Klancher

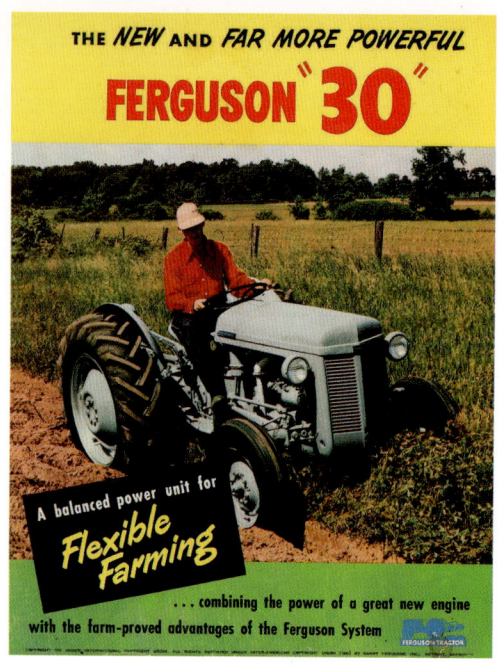

The Ferguson 30
After Ford and Ferguson parted, Harry Ferguson's company continued to produce farm tractors. The TE-20 was the first to come out after the split with Ford and was introduced in 1948. The TE-20 bore many of the improvements Harry fought for on the revised N Series Ford, including an overhead-valve engine and a four-speed transmission. The TO-30 was a refined version of the TE-20. *Voyageur Press Archives*

Funk Conversions
As power farming came to the farm, the need for more horsepower increased faster than tractor makers could update their engines and powertrains. The result was aftermarket companies that offered high-horsepower conversions for 1950s tractors. Funk was the best-known maker of engine upgrade kits for N Series Fords. The kits allowed owners to install Ford inline six-cylinder as well as V-8 engines. *Chester Peterson Jr.*

Flathead N
The Funk conversion was available as a kit that the farmer could install, and similar kits are available today. You can purchase kits to convert your N Series tractor to flathead V-8 power for about $1,000. The hood has to be widened to accommodate a larger radiator. *Chester Peterson Jr. (main)/Voyageur Press Archives (inset)*

Funk V-8-powered 8N

Funk Brothers Aviation distributed the Funk conversion kits and was owned by two brothers, Joe and Howard Funk. The kits were developed by a Nebraska farmer, Delbert Hueusinkveldt, who was inspired to make the tractor commercially available when he saw Quinton Nilson win a plow competition with a homebuilt V-8-powered 9N. *Chester Peterson Jr.*

Track Attachments
One of the interesting aftermarket options for Ford and Ferguson tractors was the addition of a track system that ran on top of the rear tires. Bombardier produced this version, and it was sold through Harry Ferguson Inc. of Detroit, Michigan. This flyer was produced in 1952.
Voyageur Press Archives

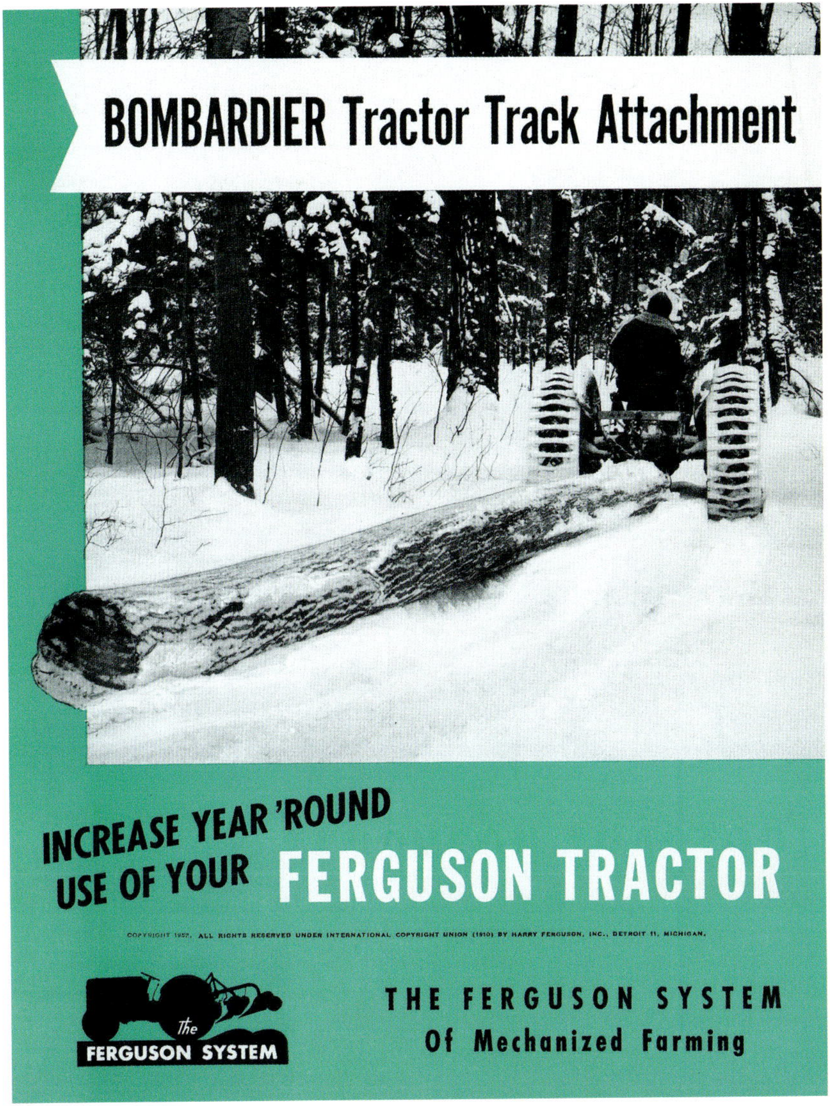

BOMBARDIER Tractor Track Attachment

INCREASE YEAR 'ROUND USE OF YOUR FERGUSON TRACTOR

COPYRIGHT 1957. ALL RIGHTS RESERVED UNDER INTERNATIONAL COPYRIGHT UNION (1910) BY HARRY FERGUSON, INC., DETROIT 11, MICHIGAN.

THE FERGUSON SYSTEM
Of Mechanized Farming

The N Series Reign

Although the N Series didn't allow Ford to wrest the number one position in the market from International Harvester, the tractors did dominate sales of small tractors and Ford slid into the number two position in the market for a while. The last of the 8N tractors were built in 1953. *Chester Peterson Jr.*

Badges

If you don't catch the twin exhaust pipes, V-8 engine, and widened hood of a Funk V-8, this little badge should clue you in to the fact that more horsepower is under the hood. The Funk brothers developed a good relationship with Ford, and Funk-converted V-8-powered tractors were sold through Ford tractor dealerships. *Chester Peterson Jr.*

Ford NAA Jubilee
Ford celebrated its 50th anniversary in 1953, and its updated model was named the Golden Jubilee to commemorate the event. The Jubilee was equipped with a 134-cubic-inch Red Tiger engine, live hydraulics, and the distinctive medallion in the hood. This Ford was photographed on the Steinberg farm in 1961. *Hasco Photographic Studio/ Minnesota Historical Society*

Fordson Major
The Fordson name lived on at the Ford plant in Dagenham, England. The tractor was known as the Ford E27N when sold in Europe, and the American import version was called the Fordson Major. The English-built tractors were available with Ford four-cylinder gas engines and Perkins six-cylinder diesel engines. *Voyageur Press Archives*

Ford Tricycles
In the 1950s, Ford Motor Company management reluctantly accepted the fact that the company needed to offer a tricycle tractor in order to gain market share. This was a departure from Henry Ford's philosophy—he believed that a wide-front tractor could do all the necessary farm tasks. The company introduced two narrow-front models, the 700 Series and 900 Series, in 1954. These were accompanied by the 600 and 800 Series, which were similar tractors with the traditional Ford wide front end. *Voyageur Press Archives*

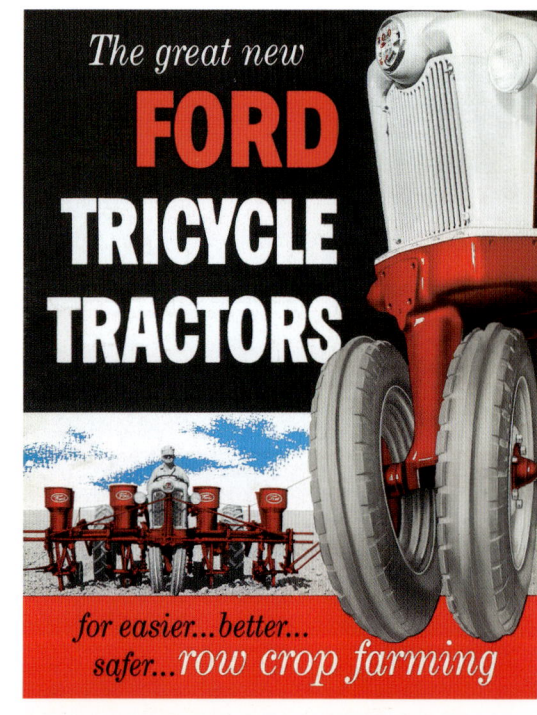

Ford 601
The N Series tractors didn't go away—they were simply refined and updated. This Model 601 was built from 1957 to 1961, and it carried a number of improvements, including the new "Red Tiger" overhead-valve engine. *Hans Halberstadt*

Ford 971
The 971 was made between 1958 and 1962 and offered the narrow-front option. The tractor was equipped with an eight-speed transmission coupled to either a gas, an LP, or a diesel Ford four-cylinder engine.
Chester Peterson Jr.

Ford 2000

The 601 Series was replaced with the Ford 2000 for the 1962 model year, and the entire line of Ford tractors was painted blue. The 2000 was available with gas or diesel engines, several different transmissions, and a live power takeoff. *Chester Peterson Jr.*

Ford 4000

The Ford 4000 replaced the 801 Series and was available in both wide- and narrow-front versions. Power was provided by gas, LP, or diesel Ford engines, and the tractor produced 46 PTO horsepower. *Minnesota Historical Society*

Ford 6000

The 6000 was the largest of the new-for-1962 model line. Ford would eventually merge with New Holland, and that company was later swallowed by Fiat. The Ford name was dropped from farm equipment in 2000. *Chester Peterson Jr.*

Chapter 5
ALLIS-CHALMERS

Allis-Chalmers was one of the most innovative tractor companies of its time when it formed in 1914, and A-C became the third-largest tractor manufacturer by 1938. Opportunistic and forward-thinking through the 1940s, the Milwaukee, Wisconsin–based company pioneered the use of pneumatic tires and was the first to experiment with a fuel cell–powered tractor in the late 1950s. The company continued successfully through the 1960s and 1970s with a complete line of farm tractors and a popular crawler line. In the 1980s, the company struggled to maintain profits. Allis-Chalmers was absorbed by Deutz-Allis in 1985. That company became part of AGCO in 1990.

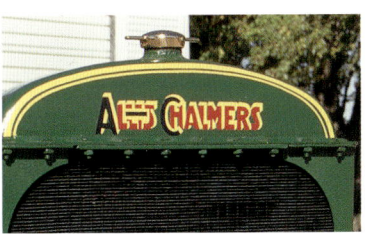

Model E 20-35

In the 1920s, Allis-Chalmers sold several farm tractors. One of its key offerings was this relatively powerful Model E 20-35, which A-C introduced in 1918 as the E 18-30 and built until 1930. Sales of the machine were fairly modest, with roughly 1,000 tractors sold each year. *Chester Peterson Jr.*

Model 10-18

The founders of the Allis-Chalmers company were Charles Allis; his chief engineer, Edwin Reynolds; and W. J. Chalmers. The founders merged four companies to create the Allis-Chalmers Company in 1901. Allis' early offerings included the 10-18, a three-wheeled lightweight machine. The company claimed the machine could list and plant 1.5 acres per hour, at a cost of 35 cents per acre. *Voyageur Press Archives*

Model L 12-20

A lighter standard tractor, the Model L 12-20 was introduced for the 1921 model year and built through 1927 (the tractor was sold as the Model L 15-25 in later production years). This tractor didn't sell terribly well, with fewer than 2,000 units sold during the eight years of production. The Allis-Chalmers tractor division was quite successful in the later 1920s, with the company adding new distribution outlets and acquiring Monarch tractors, a crawler manufacturer, and the La Crosse Plow Company. *Chester Peterson Jr.*

Filling Gaps
When Henry Ford pulled Fordson production from the United States late in the 1920s, a number of tractor parts suppliers found themselves with manufacturing equipment sitting idle. About 40 of them banded together to form the United Tractor Company and began manufacturing the United Tractor. Allis-Chalmers badged some of these machines as the Model U and sold them as Allis-Chalmers machines. United Tractor fell apart in a few years, and Allis-Chalmers continued to produce the Model U until 1952. *Chester Peterson Jr.*

Speed Demon
The Model U played a key role in history, as it was the flagship that carried the first pneumatic tires used on row-crop farm tractors. Firestone developed the tires and staged tractor races at county and state fairs. Allis-Chalmers and Firestone also staged a Milwaukee-to-Chicago "speed run." The Model U was the first Allis-Chalmers to carry the distinctive Persian Orange paint. *Voyageur Press Archives*

Race Cred
Race car driver Barney Oldfield was hired to set a new tractor speed record with a modified Model U equipped with Firestone tires. The tractor was driven to 64.28 miles per hour, and the results were highly publicized. Pneumatic tires allowed increased speeds in the field due to an improved ride, and they also improved fuel economy. *Voyageur Press Archives*

Advance-Rumely Acquisition
The successful agricultural companies of the 1920s survived by acquiring other companies in order to add new products and distribution outlets. Allis-Chalmers brought in a large dealership network and a prestigious tractor brand when it purchased Advance-Rumely in 1931. Rumely was known for building large, expensive machines with a good reputation for quality and reliability.
Voyageur Press Archives

Allis-Chalmers Rumely 6A
Allis-Chalmers became the fourth-largest tractor manufacturer in 1931. The acquisitions of Rumely and Monarch contributed greatly to this enviable position, and Ford's exit from the tractor market cleared the way for Allis-Chalmers to excel. The Model 6A was something that Rumely had developed before selling out, and it was quite similar in size and power output to the 25-40. Allis-Chalmers built this model for only two years, 1930 and 1931, and total production was fewer than 1,000 units.
Voyageur Press Archives

The A-C RUMELY "6A" TRACTOR
Steel Wheels or Air Tires

➤ Full four plow.

➤ A modern tractor for field or belt work. Handles a 32-inch grain thresher and other belt driven machinery in proportion. Power enough to haul largest combine.

➤ Belt pulley on right-hand side. Ample belt clearance. Positive and easy belt control from driver's seat. Smooth, steady power.

➤ Six speeds forward—2.50 to 4.75 miles per hour—two reverse.

➤ Heavy duty, six-cylinder engine; full force feed lubrication; efficient cooling; inserted valve seats; cut steel, hardened gears; anti-friction bearings throughout; unit construction. Can be equipped with air tires if desired.

MODEL "6A" TRACTOR WITH AIR TIRES

➤ A tractor that will give years of satisfactory service. Noted for its smooth, effortless power. Light, well balanced, easy to ride and handle. Does not overheat. Economical in use of fuel and oil.

➤ Horsepower—27, drawbar; 43, belt. Weight, 5,800 pounds.

Write for catalog giving complete details.

THE LINE OF QUALITY
— 10 —

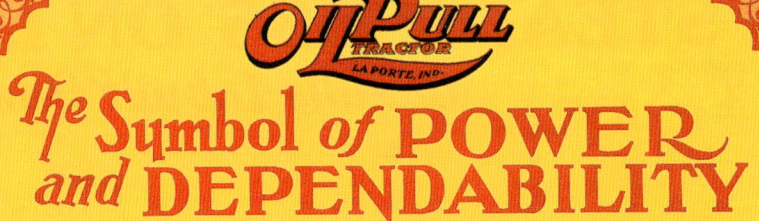

The Symbol of POWER and DEPENDABILITY

FOR nineteen years, the OilPull trade-mark has stood for undisputed excellence in tractor design, construction, and performance. Tractor users in all parts of the world have come to regard it as a symbol of power and dependability; to them it typifies—quality—sure satisfaction—extreme durability.

The first OilPulls built are still running. Their owners say they are still good for many more years of service. That service was *built into them* at the factory—that's the whole story. But there is more to it than that. Having designed and built a successful tractor to start with, Advance-Rumely engineers immediately set out to make it still better. For almost a score of years they have stuck to the original principles of OilPull design, constantly refining, improving, perfecting; with the result that the new super-powered line of OilPulls, still retaining the qualities and mechanical principles that made them famous, are today more durable, more economical and more powerful than ever before. Not only that, but many new and valuable features have been added—light weight, compact size, complete protection of working parts, ball-bearing construction, and innumerable mechanical improvements—greater value for less money.

Model U Conversion

Converting farm tractors to crawlers was done by a handful of aftermarket companies in the 1920s, and this Model U has been outfitted with a conversion kit built by the Trackson Company of Milwaukee, Wisconsin. Trackson and Allis-Chalmers were working together to produce a crawler, but those plans were waylaid when Allis bought the Monarch Tractor Corporation, which built crawlers. *Chester Peterson Jr.*

Allis-Chalmers Crawlers

In 1928, Allis-Chalmers acquired the Monarch Tractor Corporation. Monarch started out in Watertown, Wisconsin, in 1913 and later moved to Springfield, Illinois. Monarch had three sizes of crawlers, ranging from 35- to 75-horsepower machines. The 1920s were hard times for tractor makers, and Allis was able to purchase Monarch for $500,000. *Chester Peterson Jr.*

Model WC

The Model WC established Allis-Chalmers as one of the key tractor manufacturers of the early twentieth century. The row-crop tractor was introduced in 1933 and quickly became the flagship of the line. Equipped with a four-speed transmission, power lift, and a four-cylinder engine that could burn gasoline or distillate fuel, the WC was light, efficient, and popular. *Voyageur Press Archives*

Model WC

The WC was available on steel or pneumatic tires, and it sold quite well. More than 175,000 units of the model were sold between 1933 and 1948. A standard-tread version of the tractor, the WF, was built from 1937 to 1951. It was not nearly as popular as the row-crop WC. *Chester Peterson Jr.*

Model WC
The first 25 Allis-Chalmers Model WCs built used a Waukesha four-cylinder gasoline engine. An Allis-Chalmers four-cylinder engine was used on the rest of the WCs. Early engines had a 4.2:1 compression ratio, and later models had a higher 5.0:1 compression ratio. *Chester Peterson Jr.*

Model A
The big field tractor of the line, the Model A came out in 1936 as the replacement for the Model E 25-40. Powered by a 461-cubic-inch four-cylinder engine, the 7,000-plus-pound tractor was built until 1942. *Chester Peterson Jr.*

Model B

Farmers with small acreages needed tractors, and Allis-Chalmers was one of the first manufacturers to come out with a good option for them with the little B. The model was introduced for 1938 and sold for less than $500. The low price and high utility made it a popular machine, with more than 120,000 units selling between 1938 and 1957. *Chester Peterson Jr. (main)/Voyageur Press Archives (inset)*

Model RC

The RC was a blend of the B's engine in the WC chassis. The combination was underpowered and overpriced, and only about 5,500 RCs were built between 1938 and 1941. The model was replaced with the Model C in 1940. *Chester Peterson Jr.*

Freedom Rings

Farm tractor manufacturers were called on to build tanks, jeeps, and other military equipment during World War II. The material shortages created by the war left factories struggling to build enough machinery to meet demand, and Allis-Chalmers responded to that demand with this innovative used-tractor program. *Voyageur Press Archives*

FROM HIS ANVIL
Freedom Rings!

Sparks from the white-hot iron sizzled on the snow, and the clang of a hammer rang out across the frozen prairie. Out of a shapeless piece of metal he fashioned a plow point...a weapon as vital as his rifle, for without it starvation would have ambushed the caravan like an Indian massacre.

The sound of that anvil echoes today in implement repair shops across the land. It rings in the hearts of men who face unflinching an even greater emergency. For theirs is the responsibility of preparing farm machinery for the most crucial crop year in history.

Our boys in the foxholes, holding the enemy at bay, *must not have to battle hunger too.* Food must reach every continent and every village where helpless people seek the strength to unite with us in the common cause.

No other nation could begin to tackle the job. The only thing in the world that can possibly measure up to it is the efficiency of America's modern farm equipment, in the hands of people with the judgment and foresight to keep it in first-rate operating condition.

The time is growing shorter. Is your equipment Ready to Roll? It's a good feeling to see the Farm Commando emblem of honor on your A-C equipment...reconditioned by your Allis-Chalmers dealer and ready to go. An ember of resourcefulness from that little forge on the prairie flames brilliantly in your dealer's modern repair shop today.

TO BETTER LIVING
TO BETTER FARMING
TO VICTORY

**BUY WAR BONDS AND STAMPS!
INSPECT EQUIPMENT NOW!
TURN IN YOUR SCRAP!**

ALLIS-CHALMERS
TRACTOR DIVISION • MILWAUKEE • U.S.A.

FARM COMMANDO *Ready to Roll*

FARM COMMANDO *Eagle Award*
Your Allis-Chalmers dealer is awarding this beautiful red-white-and-blue Farm Commando eagle as an emblem of honor for every A-C machine inspected and pronounced ready to roll. He challenges every machine to enlist as a Farm Commando, to work overtime for your neighbor if necessary. Watch for his Farm Commando machinery and tractor school — with timely pointers from factory experts. Local officials, ag classes are welcome!

mail this **COMMANDO-GRAM**   Every Machine Has a Job to Do

Allis-Chalmers Mfg. Co., Dept. 3 Tractor Division, Milwaukee, Wisconsin
Can you help me locate the following equipment, no obligation to me:

I have the following equipment for sale to someone who needs it:

Name_____
PLEASE PRINT SIZE AND DESCRIPTION—NAME AND ADDRESS
Town_____County_____R.F.D.____State____

Model G

The postwar period saw farmers buying tractors at a pace that will never be matched again. With the economy booming and plenty of Americans still working farms, manufacturers sold as many machines as their postwar material allotments allowed them to build. One of the innovative machines brought out by Allis-Chalmers was the Model G, which was designed to work small acreages. The efficient little tractor sold well and was produced from 1948 to 1955.
Hans Halberstadt

Model WD

The WD was introduced alongside the G in 1948, and it was one of the most successful tractors produced by Allis-Chalmers. The power-adjustable rear-tread width system was useful and widely copied, and the tractor featured a live PTO and full hydraulics. Nearly 150,000 units were built and sold by 1953, the last year the WD was produced. *Chester Peterson Jr.*

Model WD-45
The replacement for the WD was the more-powerful WD-45. Introduced in 1953, the model was available with power steering and Allis' Snap Coupler hitch system. The original WD-45 had a gas engine, and a diesel engine became an available option after 1954. The model was built until 1957. *Chester Peterson Jr.*

Model D14

In the fertile tractor market of the late 1950s, Allis-Chalmers introduced two new machines, the D14 and D17, and an accompanying line of innovative implements. Farm sales had to carry the company in those days, as construction and heavy equipment sales were down. *Voyageur Press Archives*

Model D17

The D17 was produced from 1958 to 1967, with nearly 85,000 units built and sold. Six-cylinder diesel or four-cylinder gasoline or LP engines were available. Over the years, four different refinements of the D Series tractors were built, and they are known as the Series I through IV machines. The changes were mostly cosmetic. *Chester Peterson Jr.*

Model D12

In 1959, Allis-Chalmers added two new models to its line with the D10 and D12. The two tractors are nearly identical. The D10 was designed to cultivate a single row, while the D12's tread width could be adjusted so that the machine could cultivate two rows.
Lee Klancher

Model D10

This is a Series I D10, which used the 139-cubic-inch engine. Later versions were equipped with a 149-cubic-inch engine. Both were gasoline engines, and the tractor had a four-speed transmission and put out 33.5 PTO horsepower.
Chester Peterson Jr.

Model 140
The industrial version of the D10/D12 was the Industrial 40. The D10 and D12 replaced the Model CA and the Model B, respectively. A line of matching implements was offered as well. Allis-Chalmers also built an experimental fuel cell tractor in 1959. *Chester Peterson Jr.*

Model D12 High-Crop
The D12 was available in a high-clearance version. The tractor was raised with larger rear tires and longer front-axle spindles.
Chester Peterson Jr.

Model D15

The new Model D15 was introduced in 1960, along with a new All-Crop Harvester and several new combines. John Deere's introduction of its first four- and six-cylinder tractors stole a bit of thunder from Allis' new offerings, but 1960 was still a profitable year for Allis. The D15 was an upgrade and replacement for the D14, and it was produced until 1967. *Chester Peterson Jr.*

Model D21
Allis-Chalmers entered the 100-horsepower tractor market when it introduced the D21 in 1963. The 426-cubic-inch six-cylinder diesel engine made 103 PTO horsepower, giving the farmers of the day just what they wanted—big-time power. Later models were turbocharged and made 127 PTO horsepower. The model was built until 1969. *Chester Peterson Jr.*

Model 190XT
Introduced in 1964, the 190XT combined high-horsepower output with better operator comfort and improved hydraulics. Gas, LP, and diesel engines were available; all had PTO horsepower ratings of about 90. The model was built until 1971. *Chester Peterson Jr.*

Model 220

The early 1970s were not kind to most tractor manufacturers, and that was no different for Allis-Chalmers. Sales shrunk in 1971, but rebounded a bit in 1972 and 1973. One of the company's premium machines was this 220, a full-featured tractor with 136 PTO horsepower put out by a six-cylinder turbodiesel. A replacement for the D21, the 220 was built until 1973. *Chester Peterson Jr.*

Model 7080

In 1973, Allis-Chalmers introduced the 7000 Series tractors, which offered higher horsepower, more flexible transmissions, and improved operator comfort. The first of these were the 7030 and 7050. This 7080 came out in 1975, made 181 PTO horsepower, and retailed for more than $54,000. The big machine was built until 1981. *Chester Peterson Jr.*

Model 4W220
Articulated four-wheel-drive tractors were all the rage in the early 1980s, and Allis-Chalmers entered the fray with the 4W220 and 4W305. The big 305 made 305 horsepower from a twin-turbocharged six-cylinder diesel engine, while the 220 was good for 186 horsepower. Both models were built from 1982 to 1984.
Chester Peterson Jr.

Chapter 6
OLIVER

Oliver's roots are entwined with those of Henry Ford. The Oliver Chilled Plow Works made a good living building plows for Fordson tractors. When Ford neglected to develop a row-crop tractor, the Oliver family decided to pursue that opportunity. To stay alive in the tumultuous 1920s agricultural equipment market, Oliver merged with several other companies. One of those was the Hart-Parr Company, which produced a well-respected line of large tractors and had an established dealership network and factory facilities. The alliance proved a strong one, and the Oliver Farm Equipment Company that resulted produced distinctive, creatively engineered machines into the 1960s. The company was eventually absorbed by AGCO, and the Oliver name, color, and identity disappeared by the mid-1970s.

The Hart-Parr Company

One of the most historic tractor companies in America is part of the Oliver legacy. Founded by two University of Wisconsin engineering students, Charles Hart and Charles Parr, the Hart-Parr Company is credited with popularizing the term "tractor." The two friends founded the company in Madison, Wisconsin, and as its reputation and sales grew, it moved to Charles City, Iowa. In 1906, Hart-Parr's sales manager suggested that "traction engine" was too long to be used in advertisements and dubbed the machine a "tractor." *Lee Klancher*

The New Hart-Parr

The early Hart-Parr tractors were giant gas-engined beasts designed to break open prairie and supply power to threshing machines. The New Hart-Parr was one of the first "small" machines built by Hart-Parr, when the innovative owners recognized that the larger market was the mid-sized American farm. This rare example is one of seven of these machines known to exist today. *Lee Klancher*

The Oliver Chilled Plow Works
James Oliver built a successful plow manufacturing company in the mid-1800s, and his son Joseph (J. D.) continued the legacy and was one of the principals of the Oliver Chilled Plow Works. The company was an agricultural giant and built more than 750,000 plows in 1919. In 1929, Joseph led a merger of his plow company with the Nichols & Shepard Company (which made threshing machines), the American Seeding Machine Company, the Hart-Parr Company, and the McKenzie Potato Machinery Company. The company that resulted was the Oliver Farm Equipment Company. *Voyageur Press Archives*

Oliver Hart-Parr Tractors
After the 1929 merger that created the Oliver Farm Equipment Company (and absorbed Hart-Parr), the tractors Oliver produced were badged Oliver Hart-Parr tractors. One of the first of these was the row-crop Oliver Hart-Parr 18-27, which was introduced in 1930. The standard version of that tractor was the 18-28, and the more-powerful standard was the 28-44. This display of Oliver machinery took place at the Minnesota State Fair in 1934. *Minnesota Historical Society*

Oliver Hart-Parr Tractors
The 1929 merger gave Oliver a complete line of farm equipment, which was key to survival through the Great Depression. Tractors were built in Charles City, Iowa, and many of the early tractors used Waukesha engines. *Voyageur Press Archives*

Oliver 70

In the early 1930s, Oliver developed the 70. The early unstyled Model 70 is known as the Hart-Parr 70. This one at right is the later styled version, introduced in 1937. These machines were sold in Canada as the Cockshutt 70. Row-crop, standard, and orchard versions of the machine were built, and the Model 70 was produced until 1948. *Hans Halberstadt*

Oliver 70

One of the first styled tractors on the market was the Oliver 70. The model had a six-cylinder engine, power lift, and optional lights and electric start. The 70 had two companions in the line, the smaller 60 and larger 80. *Voyageur Press Archives*

Oliver Standard 80
In 1937, the old 28-44 was refined and sold as the Oliver 80. The Oliver 80 used a four-cylinder engine designed by Oliver and built by Waukesha. Both diesel and gas powerplants were available, and the model was built from 1937 to 1948. *Lee Klancher*

Cletrac Acquisition

The Cleveland Tractor Company manufactured tracked farm vehicles known as Cletracs. It incorporated in 1916 and built a complete line of crawlers ranging from the 9-horsepower Model F to the 100-horsepower Model 100. The company had about a dozen models in the lineup in 1944, when it was merged with the Oliver Farm Equipment Company to form the Oliver Corporation.
Hans Halberstadt

Cletrac Evolves

In 1946, Oliver began to paint the Cletrac models in its own color scheme and badge them as Olivers. Some of the more-popular models were the Cletrac HG and the Oliver OC-4. Oliver continued to produce a successful line of crawlers until the Oliver sale to White in 1960. *Hans Halberstadt (main)/ Voyageur Press Archives (inset)*

Oliver 66, 77, and 88

As the fires of World War II began to wane, Oliver engineers prepared for the postwar demand for tractors with a new line of machines. They introduced the Fleetline Series for 1948, with all three models sharing lots of interchangeable parts as well as live PTOs, Waukesha-built engines, and sleek, colorful sheet metal. The line was so popular that nearly 100,000 Model 77 toys sold in just three months.
Lee Klancher

Oliver 66 Orchard
The Fleetline Series tractors featured an eight-speed transmission that had six forward speeds and two in reverse. The 66 had a 126-cubic-inch gasoline or diesel four-cylinder engine and was built until 1954. *Lee Klancher*

Oliver 66 Industrial
The Oliver Corporation built a complete line of industrial equipment, including road graders, forklifts, road rollers, and power units. In 1947, the Oliver Corporation had 9,000 employees. The Fleetline Series was upgraded in 1954 with more power and features, and the new models were called the Super Series.
Lee Klancher

Oliver 99

The largest member of the Fleetline Series was based on an older design. The 99 was available with a gas or diesel engine and was rated for more than 50 PTO horsepower. The PTO was not live, and the transmission was a four-speed unit used on older models. When the 99 was updated to the Super Series in 1954, it had an all-new six-speed transmission, live PTO, and a supercharged two-cycle diesel engine.
Hans Halberstadt

Model 880

Another new line of Oliver tractors was introduced in 1958. The first models were the 550, 770, 880, and 990. Standard features included double-disc brakes, electric start, and a cigar lighter. The 880 was built until 1963. *Chester Peterson Jr.*

Model 990
The largest Oliver sported a three-cylinder two-cycle supercharged diesel engine built by General Motors. The engine put out 84 PTO horsepower and was one of the most powerful tractors on the market at the time. This chassis and engine package was sold to the Massey Ferguson Company, who put its own sheet metal and paint on it and sold it as the Massey 98. *Chester Peterson Jr.*

Model 1600

In 1960, the White Motor Company acquired the Oliver Corporation. White was a truck manufacturer, and it bought Oliver, Cockshutt, and Minneapolis-Moline in the early 1960s. The first of the new series of tractors for Oliver at the time was the 1800, which came out the year of the acquisition. Note that the red on the wheels of this 1600 is not factory original. *Chester Peterson Jr.*

Model 1600 LP High-Crop
The Oliver 1600 came out in 1962, a powerful, advanced machine that held its own in the high-horsepower dogfight between IH and John Deere. This 1962 high-crop LP version is equipped with a three-point hitch, live PTO, a Waukesha gas or diesel engine, and a Hydrospeed two-speed transmission.
Lee Klancher

Model 1900

The 1900 was also introduced in 1962 as a replacement for the 990. The 89-PTO-horsepower tractor was powered by another GM two-cycle diesel engine, this one a supercharged four-cylinder. The retail price was $9,000, and the machine was built until 1964. *Chester Peterson Jr.*

Model 1750

In the mid-1960s, Oliver introduced the new 50 Series tractors. Larger engines were the big news, and the entire line got a horsepower boost. The 1750 was built from 1966 to 1969 and put out 80 PTO horsepower. Six-cylinder gas or diesel engines were available, and the optional Hydra-Power transmission featured 12 forward and 4 reverse speeds. *Chester Peterson Jr.*

Model 1850
The 1850 was built from 1964 to 1969 and could be ordered with either a Waukesha gasoline engine or a Perkins diesel good for just over 90 PTO horsepower. The tractor retailed for $8,500 in 1969. *Chester Peterson Jr.*

Model 1950
Some tuning gave the Model 1950 nearly 20 more horsepower than the 1900, up to 106 PTO horsepower. The price was up along with the power, and the tractor retailed for $12,700 in 1974, its last year of production. *Chester Peterson Jr.*

Model 2150
Oliver put its own six-cylinder diesel engine into the 2150, a 478-cubic-inch turbocharged powerplant that put out 132 PTO horsepower. The tractor sold for $13,800 in 1969, and it was one of the last true Oliver tractors.
Chester Peterson Jr.

Model 1755

In 1969, White's reorganization consolidated Oliver and Minneapolis-Moline. The 55 Series marked the beginning of jointly produced models. The 1755 was an evolution of the 1750, and it was sold as both the Oliver 1755 and the Minneapolis-Moline G-850. By the mid-1970s, White was no longer building tractors under the Oliver name.
Chester Peterson Jr.

Chapter 7
MINNEAPOLIS-MOLINE

*J*ust as Oliver owes a tip of the hat to Henry Ford, Minneapolis-Moline can thank International Harvester for helping it form. When the Universal Tractor Company hit dire straits in the mid-1920s, International bought the company's plant, and the Moline Plow Company bought Universal. The Moline company grew again in 1929 when it merged with the Minneapolis Implement Company, the Minneapolis Threshing Machine Company, and Minneapolis Steel & Machinery. Minneapolis-Moline was the result, and the new company built Prairie Gold–painted farm tractors into the 1960s. The most distinctive model in the line was the UDLX Comfortractor, an unsuccessful experiment designed to combine the features of a truck and tractor. The line ended in the early 1970s when White stopped badging its machines as Minneapolis-Molines.

Twin City 12-20

The Twin City 12-20 was built from 1919 to 1926 and was a fairly high-tech tractor. The four-cylinder engine had four valves per cylinder, and the frame used a "unit frame" design, which meant that the engine and transmission housing were stressed frame members, which reduces production costs and overall weight. *Chester Peterson Jr. (main)/Voyageur Press Archives (inset)*

1919 Twin City 12/20

12-20 H.P., 1,000 r.p.m., 4-cylinder vertical engine mounted longitudinally. Ignition, high-tension magneto with impulse starter; cast-iron frame, unit construction; sliding-gear transmission; final drive, enclosed spur gear; belt pulley side mounted, 650 r.p.m.; forward speed, 2.2 to 2.9 m.p.h.

Twin City 20-35
The Twin City line included a number of larger tractors, and the 20-35 was one of the later, midsized offerings. Built from 1920 to 1926, the tractor was equipped with a 641-cubic-inch four-cylinder engine and an optional high-compression cylinder head. *Minnesota Historical Society*

Moline Universal
The Universal Tractor Company built the gangly front-wheel-drive tractor featured in this advertisement. The design was effective and progressive, and the rights to the name were purchased by the Moline Plow Company in 1915. Moline refined the design and installed a four-cylinder engine and electric start. Later versions had overhead-valve engines. This advertisement touts the tractor as one "that fits most farms." *Voyageur Press Archives*

The Merger

In 1929, agricultural equipment companies needed to offer a complete line of tractors and implements in order to survive in the competitive market of the times. The Minneapolis Moline Power Implement Company was formed in 1929 when the Minneapolis Steel & Machinery Company, Minneapolis Implement Company, and Minneapolis Threshing Machine Company joined forces. *Chester Peterson Jr.*

Twin City Universal JT

After the 1929 merger, the tractors produced by Minneapolis-Moline were badged with both the Twin City and the M-M names. The Model JT used a Waukesha engine, an unusual choice for a Minneapolis-Moline tractor. A conversion kit was offered to retrofit a Minneapolis-Moline engine. *Chester Peterson Jr.*

Minneapolis Steel & Machinery Company

The Minneapolis Steel & Machinery Company was organized in 1902. The company created its first tractor not long after and built large tractors that sold on the open market and to customers like J. I. Case and Deere & Company. The tractors sold on the open market were badged "Twin City Tractors." *Minnesota Historical Society*

Combines

The merger gave Minneapolis-Moline a complete line of farm tractors, including this Minneapolis-Moline combine being pulled by a Model MT tractor. Two of the companies in the merger offered combines, and the better-known machines survived in the Minneapolis-Moline line. *Minnesota Historical Society*

Twin City MTA
Built from 1934 to 1938, this Minneapolis–Moline-badged tractor replaced the Twin City MT. It was powered by a Minneapolis–Moline-built four-cylinder engine, which was good for about 30 belt horsepower. *Chester Peterson Jr.*

Prairie Gold
By 1939, Minneapolis-Moline brought out a new line of tractors. The paint scheme was Prairie Gold, and the line included the Model Z, Model R, and Model U tractors. *Voyageur Press Archives*

Model Z
The Model Z was introduced in 1937, and variants were produced until 1955. This era treated Minneapolis-Moline well, and the company sold more than 750,000 tractors in 1948—a company sales record. The Z's engine was a point of pride with the company and designed for simplicity and reliability.
Chester Peterson Jr.

Model Z and Combine
The Model Z was a staple of the Minneapolis-Moline line, but the company also continued to build and sell combines. By 1950, when this photo was taken, Minneapolis-Moline was one of a few full-line agricultural equipment companies. *A. H. Jensen/Minnesota Historical Society*

UDLX Comfortractor

In 1938, Minneapolis-Moline had this unique tractor in its lineup. Designed to replace both a tractor and a car, the Comfortractor's five-speed transmission was geared to allow a top speed of about 40 miles per hour. The Comfortractor could be ordered with a Philco radio, horn, cigar lighter, heater, gauges, and lights. Despite an extensive ad campaign, the tractor never caught on, and sales of the UDLX were minimal. These rare machines are extremely valuable today. *Chester Peterson Jr.*

Comfortractor

With room to seat three and one of the first tractor cabs, the Comfortractor had a progressive design. The tractor failed because the execution was flawed. The $1,900 retail price was double that of a typical farm tractor or a car—meaning you could purchase a tractor and a car for the same price. Also, the cab was hot and noisy inside, and the machine was difficult to drive. Fewer than 150 of them were sold. A few were sold as roadsters (without the cab). *Chester Peterson Jr. (main)/ Minnesota Historical Society (inset)*

Model U

(Previous pages)
The Model U was a key part of the Minneapolis-Moline line. This Model UB on the left is equipped with an LP gas-burning engine and a UTU is on the right. Minneapolis-Moline pioneered the use of LP gas-burning engines in 1939. *Chester Peterson Jr.*

Wartime Minneapolis-Moline

Many of the tractor manufacturers pushed the ease of operation of their machines during World War II in order to market to the farms being run by soldiers' families. Minneapolis-Moline converted some of its tractors for wartime use. *Minnesota Historical Society*

Minneapolis-Moline Z
The Model Z is one of the most prolific Minneapolis-Moline models. It was available with wide, narrow, and single-wheel front ends, in industrial and orchard versions, and with LP gas engines. Later examples could be ordered with lights, an electric starter, and a battery.
Lee Klancher (main)/Voyageur Press Archives (inset)

Minneapolis-Moline ZA
The standard version of the Minneapolis-Moline is designed for breaking new ground, and it is one of the later examples of the Model Z, using a 206-cubic-inch gasoline engine. *Lee Klancher*

Minneapolis-Moline U
The Model U was produced in a wide variety of configurations, including as a grader, a mail carrier, and an industrial version equipped with a transmission that provided six reverse and six forward speeds. *Chester Peterson Jr.*

Farm Toys
The rise of the farm tractor was paralleled with the advent of farm toys. The first mass-produced toys were sold just after the turn of the century, and by the mid-1950s, incredibly detailed replicas of the most popular farm tractors were available.
Minnesota Historical Society

Minneapolis-Moline 445

The Powerline tractors came out in the mid-1950s and included the Model 335, 445, and (later) the 550 (or Five-Star). Built from 1956 to 1959, the 445 was available with a gasoline or diesel engine. *Chester Peterson Jr.*

Minneapolis-Moline G704

In 1963, Minneapolis-Moline offered lines of farm tractors, garden tractors, industrial equipment, electronics for aviation, and custom tool and die engineering. That year the entire operation was purchased by the White Motor Company for $42 million. This LP gas front-wheel-assist model is a rare example of the G704, a model made only in 1962.
Lee Klancher

Minneapolis-Moline G708

The G700 Series of Minneapolis-Moline tractors were 100-plus-horsepower machines powered by six-cylinder 504-cubic-inch engines. Two- and four-wheel-drive versions were available, and the model came powered by gas, diesel, and LP gas engines. This 1965 front-wheel-assist G708 is one of only 31 propane-powered versions of the model built. Like the 704, the 708 was built only in 1965.
Lee Klancher

Minneapolis-Moline G1050
The White Motor Company badges adorn this 1050, the updated version of the G700 machines. This was one of the unified models and was mechanically the same as the Oliver Model 2055. The machine was built at the Minneapolis-Moline factory on Lake Street in Minneapolis. *Chester Peterson Jr.*

Minneapolis-Moline A4T-1600
This high-horsepower articulated four-wheel-drive tractor was designed under the White Motor Company banner, and the design was completed in eight months. A variety of engine options were available with horsepower ratings from 169 to 225. Retail cost for these machines was between $12,000 and $15,000. *Chester Peterson Jr.*

Chapter 8
CASE

Case, one of the pioneers of the agricultural implement industry, was founded by Jerome Increase Case in 1842 in Racine, Wisconsin, when he built and sold an innovative thresher. The J. I. Case Company grew and became a respected steam engine manufacturer. Case also built some of the first gas-engined farm tractors before the turn of the twentieth century. The "Crossmotor" Case tractors were some of the first successful small tractors, and Case went on to produce popular lines of machines through the post–World War II boom. In 1984, Case merged with parts of the International Harvester agricultural division to become Case IH. In 1999, Case IH became part of CNH Global, a giant corporation that has dealerships in 160 countries.

J. I. Case Steam Tractors

The first Case steam engine was built in about 1876. The 10-horsepower engine was on wheels and designed to be towed by a team of horses to power threshers and other farm equipment. By 1884, Case offered a self-propelled steam engine. The company's steam engine line grew to include everything from 6- to 150-horsepower machines. *Voyageur Press Archives*

Case 20-40 Gas-Kerosene Tractor

Case was one of the first manufacturers to experiment with gas tractors, having an experimental one built in 1894. The early tractors weren't effective, and the gasoline endeavor was set aside until 1910. Case contracted with the Minneapolis Steel & Machinery Company to build its first gas tractor, the Model 30-60. The 20-40 shown at left proved a more successful tractor and was sold between 1912 and 1919. *Voyageur Press Archives*

Case Farm Tractors
After a few mostly unsuccessful experiments with lighter tractors, Case hit a winning formula in 1915 with the Model 9-18. All of the models that shared the basic design of the Model 9-18 became known as the "Crossmotor" Case tractors. The models used a transverse-mounted four-cylinder engine, a one-piece cast-iron frame, and a short wheelbase. The distinctive line of machines was built until the mid-1920s. This brochure cover is from the early 1920s.
Hans Halberstadt

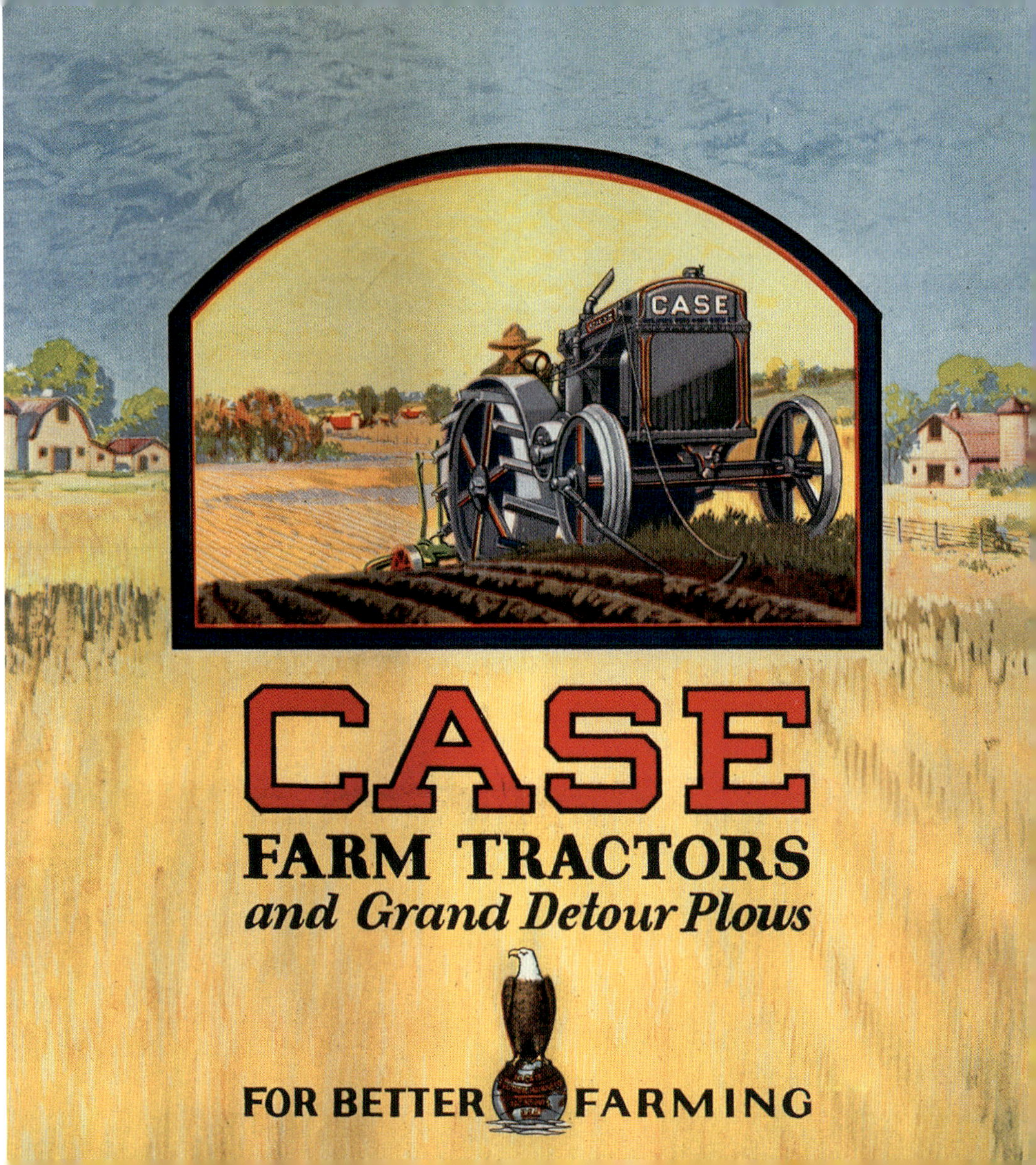

CASE

FARM TRACTORS
and Grand Detour Plows

FOR BETTER FARMING

Case Crossmotor

The Crossmotor design evolved steadily from its introduction in 1915, as power output increased in all the models and Case updated the engine with more-advanced technology. Industrial versions of the tractors were equipped with rubber-lugged steel wheels.
Hans Halberstadt

Case Crossmotor

The Crossmotor was the flagship of the Case tractor line, and that was reflected by the strong sales of the model. About 27,000 units of the smaller Crossmotor models were built and sold by the time the line was discontinued.
Hans Halberstadt

Marching Against the Beat

The last of the Crossmotors was introduced in the early 1920s. While most of the industry was downsizing, Case came out with a larger version of the Crossmotor. The 40-72 weighed more than 22,000 pounds and produced about 90 belt horsepower. The monster machine was not a big seller. *Hans Halberstadt*

Hand-Starting

Most farm tractors from the 1930s and earlier require that you turn a crank or a wheel to start them. Farmers would occasionally break an arm when the tractor backfired. The first tractors had large wheels to turn that required more arm strength and the right technique to make them work.
Hans Halberstadt

Model RC

The Model RC was introduced in 1935 as a row-crop competitor to the Farmall. The engine was a Waukesha four-cylinder. The tractor was produced until 1940, and nearly 16,000 units were built and sold.
Hans Halberstadt

Styling Case

In 1939, Case jumped on the styling bandwagon. The tractor line received sleek new sheet metal and Flambeau Red paint. The Model RC produced 17 horsepower at the belt and was sold with three- and four-speed transmissions. *Hans Halberstadt*

Chicken Roost Steering

One of the distinctive features on Case tractors is the protruding steering arm, dubbed the "Chicken Roost" steering system. This appeared on the Model CC and other Case tractor models. *Hans Halberstadt*

Case LA
A powerful tractor designed for plowing and other horsepower-heavy farm tasks, the Model LA was built from 1940 to 1952. Power was supplied by a 403-cubic-inch four-cylinder engine, and the tractor put out 58 horsepower at the belt.
Hans Halberstadt

Case LA
In 1942, Case offered Hesselman fuel-injection systems for its Model L line. This fuel delivery system allowed the Case engine to burn almost any grade of diesel fuel and was not a particularly successful option.
Hans Halberstadt

Case DO

The D Series was built from 1939 to 1953 as an upgrade of the old C Series machines. The machines were steadily upgraded throughout their production life, receiving hydraulic lifts; high-compression, diesel, and LP engines; and the Eagle Hitch, Case's proprietary hitch system. *Hans Halberstadt*

Centennial Plow

Case heavily promoted its Centennial Plow, which the company claimed was better-built and easier to use than the competition. Case also advocated the plow as an accessory that would help with soil conservation, a hot-button farming issue in the 1950s. *Hans Halberstadt*

A New Line

In 1953, Case introduced a new line of tractors and deviated from the two-letter convention of naming its machines. The line used a three-digit convention, and the first of the machines was the diesel Model 500. The 55-PTO horsepower 500 was powered by the company's first diesel engine and was the first six-cylinder produced at the Racine tractor plant. It also sported a new paint scheme.
Chester Peterson Jr.

Case 400 D

The replacement for the Model LA was the 400, which was introduced in 1955 and built until 1958. The diesel engine put out 50 belt horsepower, and the tractor's retail price was $4,400 in 1958. *Hans Halberstadt*

Case 730 Orchard

The Case 730 was built from 1960 to 1969, and the orchard versions are fairly rare. This 1966 LP gas–burning model was originally shipped to Winterhaven, Florida, to be used in the orange groves. The tractor is equipped with a hand clutch, dual-range transmission, and turf tires. *Lee Klancher*

Case 830 HC
The Case 830 featured an eight forward– and two reverse–speed transmission coupled to a gasoline, a diesel, or an LP engine, any of which was good for about 64 PTO horsepower. The five-plow tractor was introduced in 1960 and built until 1969. *Chester Peterson Jr.*

Case 1030
By the mid-1960s, if you didn't have a 100-horsepower tractor in your lineup, you weren't in the tractor-making business. The Case addition was the 1030, which produced 102 PTO horsepower and sold for just over $10,000 in 1969. *Chester Peterson Jr.*

Case IH Magnum 305

When Tenneco merged Case and International Harvester in 1984, one of the first things management wanted to do was create a brand-new machine. The first of those was the Case IH Magnum Series of tractors, which came out in 1988. The line was redesigned for 2007 and won the Good Design Award from the Chicago Athenaeum: Museum of Architecture and Design. Gold paint and special logos were applied to one hundred 305s and fifty 535s for the 20th anniversary of the Magnum line in 2008, and only select dealerships received the machines.
Lee Klancher

Chapter 9
MASSEY

Canadian businessman Hart Massey had his fingers in a lot of industrial efforts, and his entrepreneurial ways led to the creation of the Massey-Harris agricultural equipment company, which owned about half of the business in Canada by 1891. Hart's heirs brought that company into the tractor business in about 1910 by selling a variety of machines under the Massey-Harris name. The first tractor the company developed internally was the innovative four-wheel-drive Model GP in 1930, and Massey-Harris went on to develop the gorgeous Model 101 in 1939 and the useful little Pony in 1948.

Massey-Harris grew strongly in the post–World War II boom, but lost ground in the early 1950s. The tractor manufacturer allied itself with Harry Ferguson in 1953, and Massey-Ferguson was created. The line eventually was absorbed into AGCO.

Canadian Roots
Daniel Massey started building farm implements out of a blacksmith shop in Ontario in 1847. His son, Hart, took over the business and built a thriving agricultural equipment company. Alanson Harris built a similar business, and Massey and Harris battled over the agricultural equipment business in Canada. The two merged in 1891 as the Massey-Harris Company. *Voyageur Press Archives*

Massey-Harris GP
Massey-Harris sold a number of tractors before 1930, most of them machines the company purchased from other manufacturers and branded as its own. Massey-Harris purchased the Wallis line to get a jump-start on tractor manufacturing, and the first machine designed in-house was the revolutionary four-wheel-drive GP. *Hans Halberstadt*

Massey-Harris 101

In 1938, the quintessential Massey-Harris, the Model 101, appeared. It was powered by a smooth six-cylinder Chrysler engine and was covered in louvered sheet metal. The 101 featured electric start and was available in row-crop and standard-tread versions. A four-cylinder 201 was introduced in 1939. *Voyageur Press Archives*

Massey Harris 1948

The big news in 1948 was the introduction of the Pony, a one-plow tractor designed for small-acreage farms. The 62-cubic-inch four-cylinder was quiet and powerful for the tractor's size and proved a reliable piece. Massey-Harris also offered a complete line of implements made to fit the Pony. *Voyageur Press Archives*

Massey-Harris 1950 Line
At the midway point of the twentieth century, Massey-Harris had a solid reputation for powerful if less-than-cutting-edge machinery. The line included the Model 20, 30, 44, and 55. The last addition, the Pony, was the most popular model Massey-Harris built and sold during this era. *Voyageur Press Archives*

Massey-Harris 44
The Model 44 was built from 1947 to 1955 in the Racine, Wisconsin, plant. About 84,000 of these were built and sold, and the tractor was available with a variety of engines, including Continental four- and six-cylinder gasoline engines, as well as a Massey four-cylinder gasoline engine. Diesel and LP engines were offered as well. *Hans Halberstadt*

Massey Power
The Model 44 was the first Massey tractor offered with a live PTO, meaning that the power takeoff ran even when the clutch was engaged. This meant better delivery of power to implements, and it was an innovation that became industry standard in the mid-1950s. *Voyageur Press Archives*

The Ferguson Merger
On August 12, 1953, Harry Ferguson and Massey-Harris agreed to merge, and they created the second-largest farm machinery company in the world. The partnership did not go smoothly, as the company created two competing lines, the Ferguson and Massey-Harris-Ferguson tractors. This messy situation didn't last, and the lines were eventually combined to create Massey-Ferguson tractors. *Voyageur Press Archives*

The Hillary Expedition
Sir Edmund Hillary trekked to the South Pole with three Ferguson TE-20 tractors outfitted with tracks and special cold-weather wiring and batteries. After nearly three months, the group reached the pole on January 2, 1958. The team had to operate at altitudes up to 9,000 feet and negotiate ice ridges, crevasses, and blizzard conditions. *Voyageur Press Archives*

Massey-Ferguson 65
The Massey-Ferguson Model 65 was built from 1958 to 1964 in Coventry, England. The tractor could be equipped with a Continental gas or a Perkins diesel engine, and it produced 50 PTO horsepower. *Chester Peterson Jr.*

Massey-Ferguson 35
The Model 35 was also built in Coventry, England, and was produced from 1960 to 1965. The small tractor made about 37 PTO horsepower, and it was powered by Continental gas or Perkins diesel engines. *Chester Peterson Jr.*

Massey-Ferguson 98

The big Model 98 used a Detroit three-cylinder diesel engine that made 84 PTO horsepower. Built from 1960 to 1962, the tractor cost $8,188 in 1962. *Chester Peterson Jr.*

INDEX

Advance-Rumely Company, 20, 216
AGCO, 46, 207, 263, 383
Allis, Charles, 210
Allis-Chalmers, 20, 46, 207–261
   crawlers, 220
   crawler conversion, 218
   WWII used-tractor program, 232
   Models
      7000 Series, 258
      Allis-Chalmers Rumely 6A, 216
      Model 10-18, 210
      Model 190XT, 254
      Model 220, 256
      Model 4W220, 260
      Model 4W305, 260
      Model 7080, 258
      Model A, 226
      Model B, 228
      Model D10, 242, 244, 246
      Model D12, 242, 246
      Model D12 High-Crop, 248
      Model D14, 240
      Model D15, 250
      Model D17/D Series, 240
      Model D21, 252, 256
      Model E 20-35, 208
      Model G, 234
      Model I40, 246
      Model L 12-20, 210
      Model RC, 230
      Model U, 212, 214
      Model WC, 222, 224
      Model WD, 236
      Model WD-45, 238
American Seeding Machine Company, 266
Benjamin, Bert R., 90
Bombardier track attachments, 188
Case, 154, 156, 350–381
Case, Jerome Increase, 351
Case IH Steiger, 156
Chalmers, W. J., 210
Cletrac, 274, 276
Cleveland Tractor Company, 274
CNH Global, 351
Dain, Joe, 28
Deere & Company, 8, 25–83, 250
   John Deere Plow, 8, 26
   Models
      10 Series, 62
      Dain Tractor, 28, 30
      Model 1010 Industrial, 74
      Model 2010 High-Crop, 70
      Model 320, 64
      Model 3010, 72
      Model 4010, 72
      Model 4020, 78
      Model 4955, 80
      Model 70, 62
      Model 720 Diesel, 66
      Model 8010/8020, 76
      Model 830 Diesel, 68
      Model 9400, 82
      Model A / AO 38, 40, 54
      Model B, 42, 44, 46, 48, 54
      Model BO Lindeman, 48
      Model C, 36, 48
      Model D, 32, 34
      Model H/HWH, 56
      Model HX, 44
      Model L, 50
      Model M, 58
      Model R, 60, 68
      New Generation tractors (Models 1010, 2010, 3010, 4010), 25, 70, 72
Deere & Mansur Company, 28
Deere, Charles, 28
Deere, John, 25, 85
Deering, 8
Deutz-Allis, 207
Dreyfus, Henry, 52, 54, 66, 70
Ewing, William, 162, 164
Farm toys, 338
Ferguson, 58, 180, 182
   Model 30/TO-30, 182
   Model TE-20, 182, 392
Ferguson, Harry, 159, 170, 178, 383
Ferguson system, 174, 176
Fiat, 204
Firestone, 214

Ford Motor Company, 18, 159–204
   Automobile Plow, 162
   "Henry Ford & Son", 164
   Models
      600 and 800 Series, 194
      Ford 2000, 200
      Ford 6000, 204
      Ford 8N High-Crop, 180
      Ford Tricycles (700 and 900 Series), 194
      Funk V-8-powered 8N, 186
      Golden Jubilee, 192
      Mak-a-Tractor, 160
      Model 601, 196
      Model 971, 198
      Model 9N, 159, 172, 174, 176, 182
      N Series, 178, 184, 190
Ford, Henry, 86, 159, 166, 172, 178, 263
Ford, Paul B., 162
Fordson, 7, 22, 30, 34, 58, 90, 159, 164
   European, 166, 168
   Fordson Major (E27N), 194
   in Canada, 168
Froelich, John, 14
Funk, Joe and Howard, 186
Funk Brothers Aviation, 186
   badges, 190
   Funk conversions, 182, 184
Harris, Alanson, 384
Hart, Charles, 264
Hart-Parr Company, 263, 264, 266
   New Hart-Parr, 264

Hillary, Sir Edmund, 392
Hueusinkveldt, Delbert, 186
International Harvester Company (IHC), 12, 18, 20, 46, 85–157, 164, 292, 307
   Models
      Farmall 15-30
      Farmall 300 Super High-Crop, 128
      Farmall 400, 126
      Farmall A, 108
      Farmall B, 108, 116
      Farmall C, 116
      Farmall Cub, 114, 142
      Farmall F-12, 94, 98
      Farmall F-30, 92, 98, 174
      Farmall H, 85, (HV) 106, 128
      Farmall M, 85, 104, 110, 112
      Farmall Regular, 85, 90
      Farmall Super C, 120
      Farmall Super M-TA, 122, 126
      Farmall W-12, 98
      Farmall W-30, 98
      International 284, 152
      International 340, 140
      International 350 High-Crop LP, 134
      International 450 with Cummins Diesel, 136
      International 560 Diesel, 138
      International 600, 130
      International 706, 146
      International 1256, 148
      International 1468, 150
      International 2806, 144
      International 4366, 156

      International 5288, 154
      International Four Series, 110
      International I-12
      International T-6 crawler, 112
      International TD-14A, 118
      International W-6 and Six Series, 110, 112
      International Wheatland 350, 132
      McCormick-Deering 10-20, 90
      McCormick-Deering 15-30, 90, 104
      TracTracTor, 100
      Type A, 12
J. I. Case, 8
J. I. Case Steam Tractors, 352–353
John Deere, 25–83, 292
   and Waterloo, 14
La Crosse Plow Company, 210
Lindeman, Jesse, 48
Lindeman Company, 48
Loewy, Raymond, 52, 104, 106, 110, 112
Mansur, Alvah, 28
Massey, Daniel, 384
Massey, Hart, 383–384
Massey Ferguson Company, 288, 383–399
   Model 35, 394
   Model 65, 394
   Model 98, 396
   Model 990, 288
Massey-Harris Company, 46, 383, 392
   1950 line, 388
   Model 44, 388–390
   Model 101, 386
   Model GP, 384

Pony, 386
McCormick, 8
McCormick, Cyrus, 85, 86
McCormick, Nettie Fowler, 86
McKenzie Potato Machinery Company, 266
Minneapolis Steel & Machinery Company, 307, 314, 352
   Twin City Tractors, 314
Minneapolis-Moline, 46, 290, 304, 307–349
   Minneapolis Moline Power Implement Company
     merger, 312
   Wartime conversions, 330
   Models
     Combine and Model MT, 314
     Model 708, 344–345
     Model A4T-1600, 348–349
     Model G704, 342–343
     Model G-850 (same as Oliver 1755), 304
     Model G1050, 346–347
     Model U, 325–329, 336–337
     Model Z, 320 332–33
     Model Z and combine, 322
     Model ZA, 334–335
     Powerline Model 445, 340–341
     Prairie Gold Models Z, R, and U, 318
     UDLX Comfortractor, 307, 324, 326

Twin City 12-20, 308
Twin City 20-35, 310
Twin City MTA, 316
Twin City Universal JT, 312
Minnesota Steam Tractor, 10
Moline Plow Company, 307, 310
Monarch Tractor Corporation, 210, 218, 220
New Holland, 204
Nilson, Quinton, 186
Oldfield, Barney, 214
Oliver Farm Equipment Company, 46, 263–305
   Oliver Chilled Plow Works, 263, 266
   Oliver Corporation, 274
   Models
     Fleetline Series (Models 66, 77, 88), 278, 280
     Model 880, 286
     Model 990, 288
     Model 1600, 290
     Model 1600 LP High-Crop, 292
     Model 1750, 296
     Model 1755, 304
     Model 1850, 298
     Model 1900, 294
     Model 1950, 300
     Model 2150, 302
     Oliver 66 Industrial, 282

Oliver 66 Orchard, 280
Oliver 70, 270
Oliver 80, 272
Oliver 99, 284
Oliver Hart-Parr Tractors, 266, 268
Oliver, James, 266
Oliver, Joseph (J. D.), 266
Parr, Charles, 264
Porsche, Ferdinand, 52
Reynolds, Edwin, 210
Rumely, Meinrad and Jacob, 20
Rumely OilPull, 20
Steiger, 156
Tenneco, 85, 154, 380
Trackson Company, 218
Tractor wars, 166
United Tractor Company, 212
   Moline Universal, 310
Universal Tractor Company, 307, 310
University of Nebraska, 62
Waterloo Boy Tractor, 14, 30
   Waterloo Boy N, 16, 32
Waterloo Gasoline Engine Company, 14
Waukesha engines, 96, 224, 268, 272, 278, 292, 298, 312, 362
White Motor Company, 290, 304, 307, 346
Wright, Russel, 52